Fish Passage in Large Culverts with Low Flows

FOREWORD

The study described in this report was conducted at the Federal Highway Administration's Turner-Fairbank Highway Research Center (TFHRC) J. Sterling Jones Hydraulics Laboratory in response to the need for guidance to evaluate the variation of velocity within a culvert cross-section to facilitate fish passage design identified by State transportation departments. The State transportation departments that contributed funding for the realization of this study were Alaska, Georgia, Maryland, Michigan, Minnesota, Vermont, and Wisconsin. A 3-ft corrugated metal pipe (CMP) was used during the physical modeling phase of the research study. The results obtained from the physical modeling phase of the study were corroborated during the computational fluid dynamics numerical modeling phase of the study. This report will be of interest to hydraulic engineers and environmental scientists involved in the design of new or retrofit of existing CMPs for fish passage. The report is being distributed as an electronic document through the TFHRC Web site at http://www.fhwa.dot.gov/research/.

Jorge E. Pagán-Ortiz
Director, Office of Infrastructure
Research and Development

TECHNICAL REPORT DOCUMENTATION PAGE

1. Report No. FHWA-HRT-14-064	2. Government Accession No.	3. Recipient's Catalog No.
4. Title and Subtitle Fish Passage in Large Culverts with Low Flows		5. Report Date August 2014
		6. Performing Organization Code
7. Author(s) Yuan Zhai, Amin Mohebbi, Roger Kilgore, Zhaoding Xie, and Jerry Shen		8. Performing Organization Report No.
9. Performing Organization Name and Address GENEX SYSTEMS, LLC 2 Eaton Street, Suite 603 Hampton, VA 23669		10. Work Unit No. (TRAIS)
		11. Contract or Grant No. DTFH61-11-D-00010
12. Sponsoring Agency Name and Address Office of Infrastructure Research and Development Federal Highway Administration 6300 Georgetown Pike McLean, VA 22101-2296		13. Type of Report and Period Covered Laboratory Report October 2010–December 2013
		14. Sponsoring Agency Code

15. Supplementary Notes

The Contracting Officer's Representative (COR) was Kornel Kerenyi (HRDI-50). Steven Lottes and Cezary Bojanowski of the Argonne National Laboratory and Dr. Fred Chang contributed to this research. The Maryland State Highway Administration was the lead State coordinating with FHWA on this research.

16. Abstract

A series of physical and numerical modeling runs were completed to support the development of a design procedure for characterizing the variation in velocity within non-embedded and embedded culverts. Physical modeling of symmetrical half-section circular culverts was conducted to provide data against which computational fluid dynamics (CFD) modeling could be validated. The initial CFD modeling featured two-phase numerical computations that successfully reproduced the physical modeling results. To further simplify, single-phase modeling and truncated single-phase modeling were evaluated with good results. For the embedded culvert runs, a successful strategy for representing natural bed material within the culvert was developed.

Once the CFD modeling was validated by the physical modeling, the CFD modeling was used to analyze the full culvert cross-sections. Test matrices included CFD runs scaled up to larger culvert sizes. One series of runs maintained Froude number based scaling and one series tested larger sizes without the scaling constraint. The CFD runs and a velocity distribution model formed the basis of a proposed design methodology for determining the velocity distribution within a culvert cross-section. Using the 42 CFD runs for a 3-ft diameter culvert, the 5 parameters necessary for the velocity model were estimated. Then, based on geometric and hydraulic parameters available to a designer, relations were developed to estimate those parameters. The approach was successfully validated on CFD runs for 6-ft and 8-ft diameter culvert models. The proposed design procedure allows a designer to estimate the velocity throughout a cross-section. These data may be depth-averaged to provide a distribution of velocity and depth across the culvert cross-section that may be used to evaluate fish passage. Although developed for circular culverts, the parameters used in the method are such that the procedure should be applicable to rectangular and other shapes. Two design examples and an application guide are provided to illustrate the method and the required computations.

17. Key Words culvert hydraulics, fish passage, velocity distribution, CFD, culvert embedment		18. Distribution Statement No restrictions. This document is available to the public through NTIS: National Technical Information Service 5301 Shawnee Road Alexandria, VA 22312	
19. Security Classif. (of this report) Unclassified	20. Security Classif. (of this page) Unclassified	21. No. of Pages 134	22. Price

Form DOT F 1700.7 (8-72)   **Reproduction of completed page authorized**

SI* (MODERN METRIC) CONVERSION FACTORS

APPROXIMATE CONVERSIONS TO SI UNITS

Symbol	When You Know	Multiply By	To Find	Symbol
		LENGTH		
in	inches	25.4	millimeters	mm
ft	feet	0.305	meters	m
yd	yards	0.914	meters	m
mi	miles	1.61	kilometers	km
		AREA		
in^2	square inches	645.2	square millimeters	mm^2
ft^2	square feet	0.093	square meters	m^2
yd^2	square yard	0.836	square meters	m^2
ac	acres	0.405	hectares	ha
mi^2	square miles	2.59	square kilometers	km^2
		VOLUME		
fl oz	fluid ounces	29.57	milliliters	mL
gal	gallons	3.785	liters	L
ft^3	cubic feet	0.028	cubic meters	m^3
yd^3	cubic yards	0.765	cubic meters	m^3
		NOTE: volumes greater than 1000 L shall be shown in m^3		
		MASS		
oz	ounces	28.35	grams	g
lb	pounds	0.454	kilograms	kg
T	short tons (2000 lb)	0.907	megagrams (or "metric ton")	Mg (or "t")
		TEMPERATURE (exact degrees)		
°F	Fahrenheit	5 (F-32)/9 or (F-32)/1.8	Celsius	°C
		ILLUMINATION		
fc	foot-candles	10.76	lux	lx
fl	foot-Lamberts	3.426	candela/m^2	cd/m^2
		FORCE and PRESSURE or STRESS		
lbf	poundforce	4.45	newtons	N
lbf/in^2	poundforce per square inch	6.89	kilopascals	kPa

APPROXIMATE CONVERSIONS FROM SI UNITS

Symbol	When You Know	Multiply By	To Find	Symbol
		LENGTH		
mm	millimeters	0.039	inches	in
m	meters	3.28	feet	ft
m	meters	1.09	yards	yd
km	kilometers	0.621	miles	mi
		AREA		
mm^2	square millimeters	0.0016	square inches	in^2
m^2	square meters	10.764	square feet	ft^2
m^2	square meters	1.195	square yards	yd^2
ha	hectares	2.47	acres	ac
km^2	square kilometers	0.386	square miles	mi^2
		VOLUME		
mL	milliliters	0.034	fluid ounces	fl oz
L	liters	0.264	gallons	gal
m^3	cubic meters	35.314	cubic feet	ft^3
m^3	cubic meters	1.307	cubic yards	yd^3
		MASS		
g	grams	0.035	ounces	oz
kg	kilograms	2.202	pounds	lb
Mg (or "t")	megagrams (or "metric ton")	1.103	short tons (2000 lb)	T
		TEMPERATURE (exact degrees)		
°C	Celsius	1.8C+32	Fahrenheit	°F
		ILLUMINATION		
lx	lux	0.0929	foot-candles	fc
cd/m^2	candela/m^2	0.2919	foot-Lamberts	fl
		FORCE and PRESSURE or STRESS		
N	newtons	0.225	poundforce	lbf
kPa	kilopascals	0.145	poundforce per square inch	lbf/in^2

*SI is the symbol for the International System of Units. Appropriate rounding should be made to comply with Section 4 of ASTM E380.
(Revised March 2003)

TABLE OF CONTENTS

LIST OF FIGURES

LIST OF TABLES

LIST OF ABBREVIATIONS AND SYMBOLS

2D	Two-dimensional
3D	Three-dimensional
A	Cross-section flow area, ft^2 (m^2)
ADV	Acoustic Doppler velocimetry
A_p	Surface area of the wetted perimeter (culvert wall and bed), ft^2 (m^2)
a_y	Variable dependent on embedment
B_i	Transverse distance on the water surface to the left (i = 1) and right (i =2), ft (m)
B_{avg}	Average flow width in the cross-section (half section), ft (m)
CAD	Computer-aided design
CCD	Charge coupled device
CFD	Computational fluid dynamics
CMP	Corrugated metal pipe
CBC	Concrete box culvert
CSP	Corrugated structural plate
d_e	Embedment depth, ft (m)
D	Culvert diameter, ft (m)
D_{50}	Median grain size, ft (m)
FHWA	Federal Highway Administration
Fr	Froude number based on average cross-section velocity and depth $(V_a/(gy_a)^{0.5})$
g	Gravitational acceleration, ft/s^2 (m/s^2)
HEC-RAS	Hydrologic Engineering Centers River Analysis System
M	Velocity distribution parameter
n	Manning's roughness
p	Pressure on the flow field at the end of the control volume, lb/ft^2 (N/m^2)
Δp	Change in pressure from one end of the control volume to the other, lb/ft^2 (N/m^2)
P_b	Wetted perimeter for the bed material, ft (m)
PIV	Particle image velocimetry
P_w	Wetted perimeter for the culvert wall, ft (m)
Q	Culvert discharge, ft^3/s (m^3/s)
Q_H	High passage discharge, ft^3/s (m^3/s)
Q_L	Low passage discharge, ft^3/s (m^3/s)
RANS	Reynolds-averaged Navier-Stokes
Re	Reynolds number based on hydraulic radius (V_aR_h/v)
R_h	Hydraulic radius, ft (m)
RMS	Root mean square
RMSE	Root mean square error
SPIV	Stereoscopic particle image velocimetry
T	Top width of the water surface, ft (m)
u_*	Mean shear velocity, ft/s (m/s)
V	Point velocity, ft/s (m/s)
V_a	Average flow velocity, ft/s (m/s)
V_f	Maximum allowable fish passage velocity, ft/s (m/s)
V_i	Depth average velocity of the ith cross-section slice, ft/s (m/s)
V_{max}	Maximum flow velocity in a cross-section, ft/s (m)

VOF	Volume of fluid
y	Vertical distance from the lowest elevation of the flow field, ft (m)
Y	Normalized Cartesian coordinate in the vertical direction
y_a	Average flow depth (A/T), ft (m)
y_f	Minimum required depth for fish passage, ft (m)
y_i	Flow depth of the ith cross-section slice, ft (m)
y_{max}	Maximum flow depth, ft (m)
z	Horizontal distance from the culvert centerline, ft (m)
Z	Normalized Cartesian coordinate in the horizontal direction
β_i	Velocity distribution parameter
δ_i	Velocity distribution parameter, ft (m)
δ_y	Velocity distribution parameter, ft (m)
ε	Velocity distribution parameter, ft (m)
η	Coordinate axis in the ξ-η system
ν	Kinematic viscosity, ft^2/s (m^2/s)
τ_w	Wetted perimeter shear stress, lb/ft^2 (N/m^2)
ξ	Coordinate axis in the ξ-η system
ξ_0	Minimum value for the ξ coordinate
ξ_{max}	Maximum value for the ξ coordinate

CHAPTER 1. INTRODUCTION

Historically, culverts have been designed for hydraulic efficiency without consideration of fish passage or, more generally, aquatic organism passage. Over time, it has become apparent that culverts frequently become impediments to healthy aquatic ecosystems because they can prevent the movement of fish and other aquatic organisms upstream and downstream through the culvert. Therefore, aquatic organism passage through culverts has become an important design element component for road/stream crossings. Common physical characteristics that may create barriers include high water velocity, shallow water depth, large outlet drop heights, turbulence within the culvert, and accumulation of debris. Sediment deposition and erosion at the culvert may also create a barrier to passage. Culvert length, slope, and roughness may create conditions that impede passage as well.

Further complicating design is that passage needs differ by species, life stage, and season. To address this complex task, the Federal Highway Administration (FHWA) developed a stream simulation approach for designing culverts.[1] Stream simulation is based on the concept that if conditions inside a culvert are similar to those conditions in the stream upstream and downstream of the culvert, then aquatic organism passage will be provided without consideration of the specific physical requirements of one or more species.

However, stream simulation is not appropriate for all situations. For example, an existing culvert that is blocking passage may not be a good candidate for replacement using stream simulation because of the size of the embankment or insufficient budget for a replacement. Applications of stream simulation may also be limited for new culvert installations. Site constraints or budget limits could dictate a smaller culvert installation than would be recommended by stream simulation. In these cases, it may be desirable to design a culvert crossing considering the specific passage needs of a specific species of fish. Doing so requires an understanding of the migration seasonality, life stage swimming capabilities, and stream flow rates expected during passage. Ideally, this information is developed by a multidisciplinary team of aquatic biologists, hydrologists, and engineers. From this information, the maximum velocity and minimum depth requirements for the target fish are derived.

Considering only average velocity in a culvert masks that there are zones within the flow field where velocities both higher and lower than the average exist. The objective of this research is to assist in the design of culverts for fish passage by 1) identifying zones of lower velocity that are conducive to fish passage and 2) developing practical design methods quantifying these lower-velocity zones.

In addition to this introduction, this report includes a literature review (chapter 2) and a description of the methodology for this study (chapter 3). The experimental methods are described, including the physical modeling in a test flume (chapter 4) and numerical modeling using computational fluid dynamics (CFD) (chapter 5). Chapter 6 provides development of the analytical findings and recommended design techniques. Chapter 7 provides a summary and conclusions.

CHAPTER 2. LITERATURE REVIEW

The literature review for this study included identifying fish passage requirements, general approaches to designing culverts, and laboratory and numerical studies oriented to understanding and achieving fish passage. With respect to fish passage, stream simulation is generally considered the state of the art in fish passage design. Stream simulation attempts to mimic the conditions experienced by aquatic organisms upstream and downstream of the culvert within the culvert itself. The guiding principle is that if the organism can move up and down the stream, then it can also move through the culvert. Of critical importance is that by using this approach, the designer does not need specific knowledge of the habits and capabilities of specific species. This approach has allowed broader attention to aquatic organisms of many types rather than a singular focus on a fish species and life stage. The first comprehensive guide to stream simulation was developed by the United States Forest Service.[2] Subsequently, FHWA developed its variation on stream simulation.[1] Both documents provide references supporting the use of stream simulation. An earlier synthesis report prepared for FHWA provides a good overview of previous work in culvert design for fish passage.[3] The synthesis report, when used to identify culvert design specifications, facilitates consensus and expedites permitting for project delivery.

Laboratory studies have been conducted related to culverts and other hydraulic devices such as fishways. These studies used data collection techniques and approaches that are applicable to studying culverts. For example, Puertas et al. studied vertical-slot fishways using acoustic Doppler velocimeters to measure three-dimensional (3D) velocities to capture the flow structure and velocity distribution.[4] Magura details physical modeling to investigate the flow characteristics of circular corrugated structural plate (CSP) culverts with 10-percent embedment and projecting end inlets.[5]

Numerical methods have played an important role in estimating the velocity distribution at a stream cross-section or within a culvert. (See references 6 through 11.) Mathematical and computational methods can assist in the prediction of the velocity distribution when field data are unavailable or insufficient and laboratory data are limited. Blank et al. investigated the effect of culvert velocity on fish passage with CFD for estimating the 3D velocity field and characterized energy expenditure paths to identify passageways.[10] Haque et al. validated a 3D Reynolds-averaged Navier-Stokes (RANS) model to predict flow and stratification effects related to fish passage at hydropower dams.[12] House et al. presented a regression model that estimates the percentage of a cross-section suitable for fish passage in embedded culverts based on discharge, total cross-sectional area, Froude number, and relative roughness.[9]

Numerical models, validated by field or laboratory data, enable simulation of velocity patterns and secondary flow structures in complex natural channels and other hydraulic structures. (See references 13 through 16.) Khan et al. developed and validated a 3D CFD model of a dam forebay.[17]

CHAPTER 3. METHODOLOGY AND ANALYTICAL FORMULATION

At the flood flows for which culverts are typically designed, knowing the flow depth and average velocity within the culvert and in the stream channel are generally sufficient for design at these higher discharges.[18] However, at lower flows of interest when considering fish passage, the velocity distribution within the culvert may become relevant for design. The objective of this fish passage research is to assist in the design of culverts by identifying zones of lower velocity during low flows that are conducive to fish passage and to develop practical design methods based on quantifying these lower-velocity zones. The focus of the study is corrugated metal pipes.

To accurately evaluate the ability of specific fish with specific swimming capabilities to traverse corrugated metal culverts, it is desirable to examine the velocity distribution within the culvert flow field to identify zones of lower velocity adjacent to the culvert wall under low flow conditions. Other studies have documented the tendency of fish to seek out a swimming location with the lowest velocity.[19,20]

This study addresses low flows in culverts as illustrated in figure 1. For low flows, a fraction of the total barrel opening is carrying water, and the flow depths are shallow compared with the culvert diameter. As with high flows, the low flow condition may be characterized by its discharge Q, average velocity V_a, and maximum depth y_{max}.

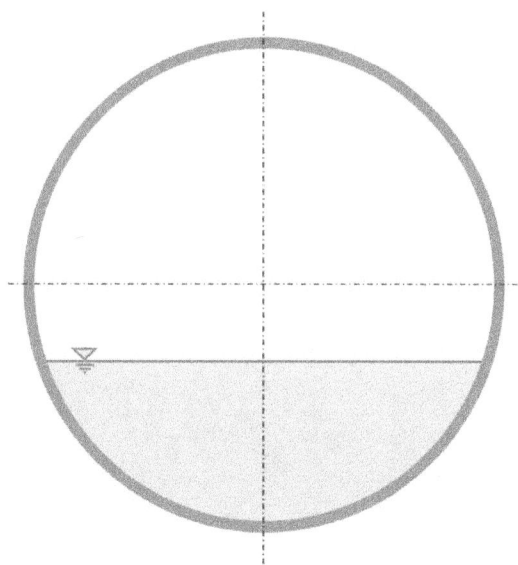

Figure 1. Illustration. Culvert water surface at low flow.

Figure 2 illustrates the variation in depth-averaged velocity and depth in a circular cross-section. Typically, the local depth-averaged velocity (V_1, V_2, V_3, etc.) will approach zero at the culvert wall and will be at a maximum near the center of the culvert cross-section. The depths y_1, y_2, y_3, etc. will also vary as shown in the figure.

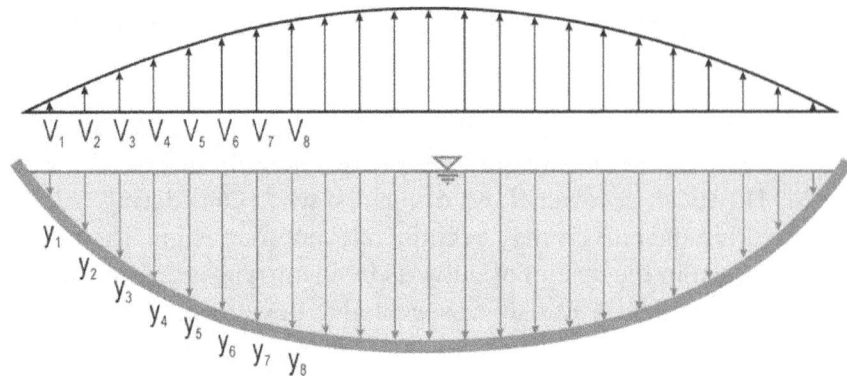

Figure 2. Illustration. Variation of velocity and depth in a circular cross-section.

Recommendations for a required minimum depth (y_f) and maximum allowable velocity (V_f) are expected to vary by location, species, and life stage. Other important considerations may include the sustained and prolonged speeds of the various species for which the fish passage is being designed. The minimum depth and maximum allowable velocity will generally be provided by a team that includes an aquatic biologist familiar with fish passage needs.

The design hypothesis is that even though the average velocity in the culvert exceeds the maximum allowable velocity for fish passage, a fish path may exist where the local velocities do not exceed the allowable velocity and provide sufficient depth (and width) for passage. Figure 3 illustrates this concept by showing a path on both the left and right sides of the symmetrical cross-section where hydraulic conditions are favorable for passage. As illustrated, the depth-averaged velocity in the eighth strip, V_8, is less than or equal to V_f, while the depth-averaged velocity in the ninth strip is greater. In addition, depth in the fish path must exceed the minimum fish depth, y_f, over a sufficient width to allow passage.

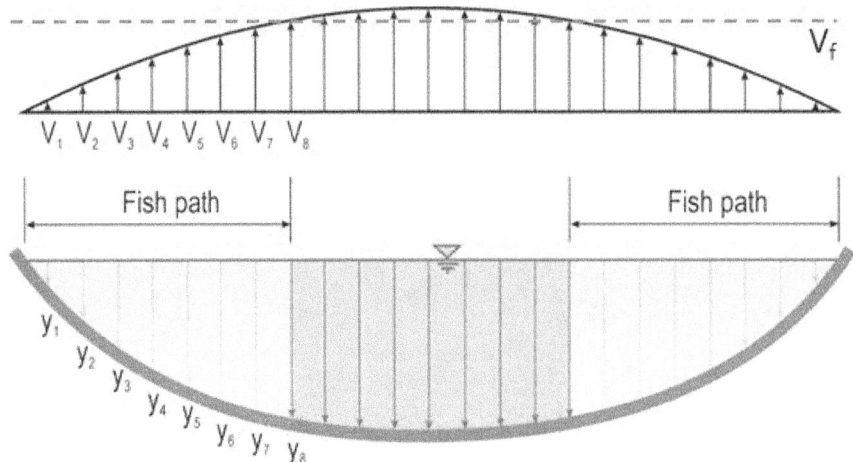

Figure 3. Illustration. Fish path concept.

The necessary fish path must exist during the periods at which the subject fish species/life stage is moving up and down stream for a successful design. *Culvert Design for Aquatic Organism Passage* recommends the use of the high and low passage flows to define the discharge range within which fish passage movement is anticipated.[1] The document also provides recommendations on estimating the high and low passage flow rates.[1]

The high passage flow, Q_H, represents the upper bound of discharge at which fish are believed to be moving within the stream. At this discharge, a fish path must exist if fish passage is to be facilitated; achieving a sufficiently low velocity is typically the challenge. The culvert discharge that includes an acceptable fish path is compared with the high passage discharge. If the culvert discharge is equal to or greater than the high passage discharge, then this criterion is satisfied. If not, then the design of the crossing must be revised until the criterion is satisfied.

Similarly, the low passage flow, Q_L, represents the lower bound of discharge at which fish are believed to be moving within the stream. At this discharge, a fish path must also exist if fish passage is to be facilitated; achieving the minimum depth is typically the challenge. The culvert discharge that includes an acceptable fish path is compared with the low passage discharge. If the culvert discharge is equal to or less than the low passage discharge, then this criterion is satisfied. If not, then the design of the crossing must be revised until the criterion is satisfied.

For the foregoing design approach to succeed, the following conditions must be met:

- The necessary design criteria for the target fish species, (i.e., maximum velocity, minimum depth, and minimum width) must be available.

- The channel width and depth at the culvert outlet should provide a flow path and a flow depth sufficiently deep to facilitate the entry of fish into the culvert at the low passage flow. The downstream and upstream culvert inverts should be at or below the elevation of the bed of the outlet and inlet channel, respectively.

- The channel width and depth upstream of the culvert should transition to the width of the culvert entrance in a manner that avoids any significant contraction, drawdown of the flow, or critical flow at the culvert entrance at both the low and high passage flows.

CHAPTER 4. PHYSICAL MODELING

INTRODUCTION

The physical flume modeling is described in this chapter. Particle image velocimetry (PIV) and acoustic Doppler velocimetry (ADV) are employed to achieve detailed representation of the flow field within the experimental apparatus. In addition to traditional PIV, stereoscopic particle image velocimetry (SPIV) is applied as a tool to characterize the 3D velocity field.

The Cartesian coordinate system for all analyses is defined by the x-axis representing the longitudinal (flow) direction. Looking downstream, the vertical direction (depth) is represented by the y-axis, and the horizontal direction (width) is represented by the z-axis.

EXPERIMENTAL SETUP

Flume

The physical experiments were carried out in a rectangular cross-section tilting flume at Turner-Fairbank Highway Research Center in McLean, VA. The flume is 30 ft long and 4 ft wide and is shown in figure 4. The culvert test section is 16.2 ft long and 1.5 ft wide. The flume can be tilted up to 3 degrees. A honeycomb was placed in the trumpet-shaped inlet to channel the water to ensure flow homogeneity. Uniform steady-state flow conditions were controlled with screw jacks (to raise the upstream end of the flume while lowering the downstream end) and a bottom-hinged flap gate in the exit section. Fifteen ultrasonic sensors measured the water surface profile. A 4 ft^3/s pump was used for circulating water into the flume. The flume was mounted by a two-dimensional (2D) robot to facilitate measurement of the velocity field.

Figure 4. Photo. Fish passage culvert experimental flume.

ADV

A 16 MHz microADV from SonTek™ was used to collect velocity measurements. It was mounted on a robot, as shown in figure 5, and controlled with a LabView™ program. These measurements were used to evaluate the accuracy and efficiency of other velocity measurement techniques and the CFD modeling. Fifteen hundred sample points were collected for each velocity measurement to precisely average fluctuating streamlines. For more information on ADV, see appendix B of this report.[21]

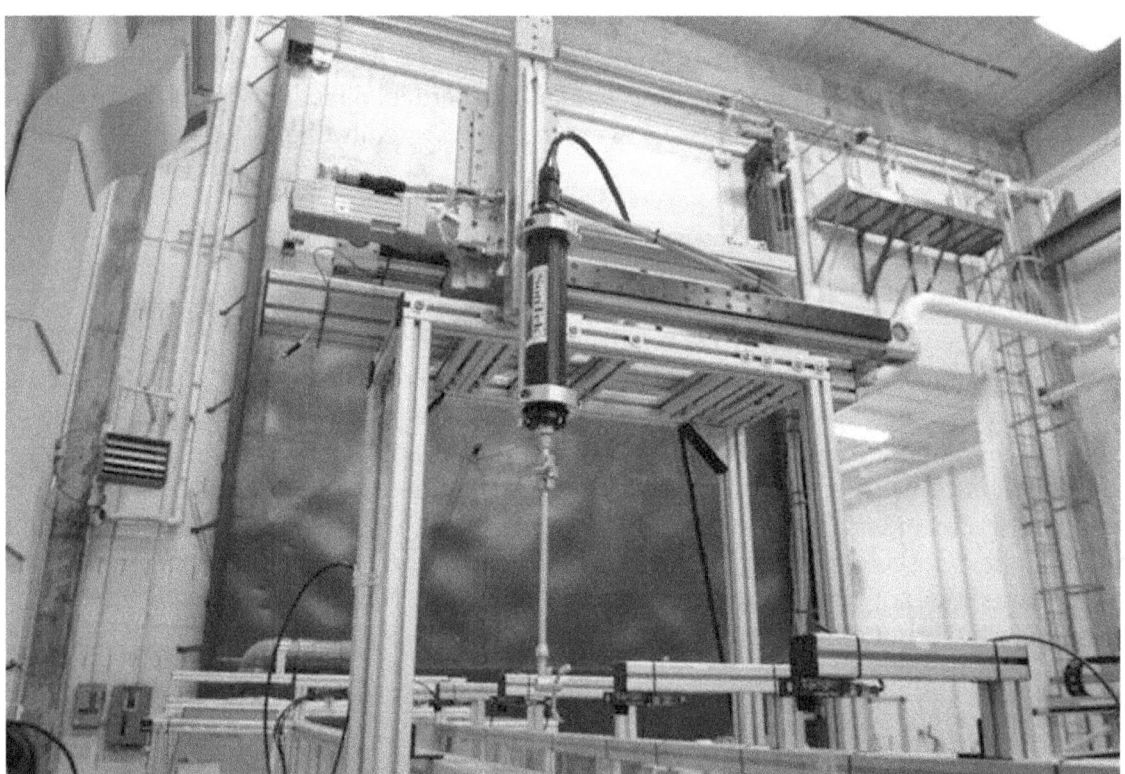

Figure 5. Photo. SonTek™ ADV mounted on the data collection robot.

PIV/SPIV

PIV and SPIV were used in these experiments to collect more comprehensive and detailed representations of the velocity field than is achievable using ADV. These methods are non-intrusive whole flow field techniques for assessment of the mean and instantaneous velocity vectors within a single plane of interest.

In SPIV, two coupled cameras capture the same plane at the same time but with different off-axis view angles. A double-pulsed Solo 120 PIV New Wave Research™ Nd:YAG laser along with a pair of Megaplus™ ES 1.0 digital cameras from Roper Scientific MASD Inc. were configured in LabView™ to operate in a synchronized manner. The SPIV setup is shown in figure 6, and a close-up of the laser is shown in figure 7.

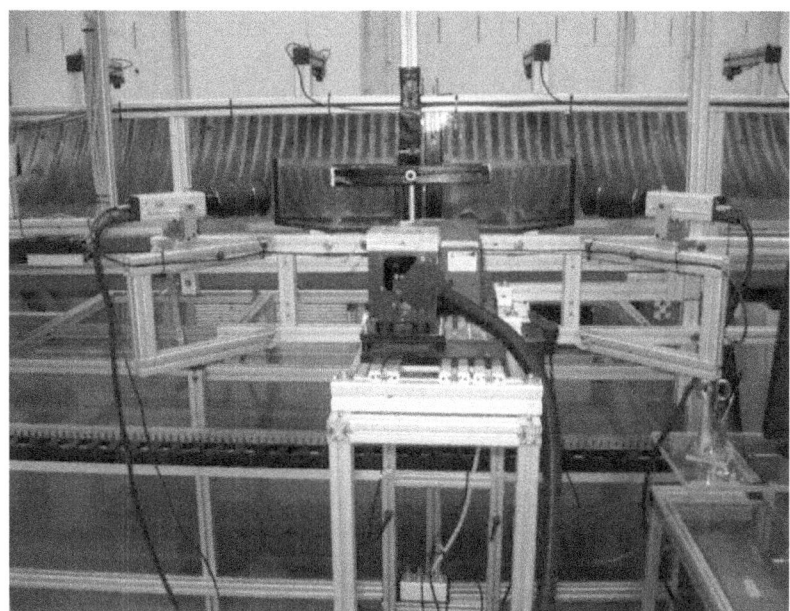

Figure 6. Photo. SPIV configuration.

Figure 7. Photo. Nd:YAG laser.

The spatial resolution was limited to 960 (horizontal) by 960 (vertical) pixels, which was implemented in the charge coupled device (CCD) array. The camera was connected through a 68-pin small computer system interface cable to the frame grabber card and could capture up to 30 images per s. It also featured a built-in electronic shutter with exposure times as short as 127 ms for maximum flexibility and performance when imaging fast moving objects. CCD cameras were equipped with two Sigma zoom lenses with a focal length ranging from 1.1 to 2.8 inches and an 8-level f-number from 2.8 to 32.

Silver-coated hollow glass spheres (AGSL150-16TRD from the Potters Industries Inc., Carlstadt, NJ) were introduced into the flow with an average diameter of 69 μm, density of 0.93 g/cm^3, and

11

17.7-percent weight of silver coated on their surfaces. For further information, see appendix B of this report.[22]

TEST MATRIX

Symmetrical half sections of corrugated metal pipe (CMP) were used to maximize the size of culvert that would fit into the flume with one of the flume walls serving as the culvert centerline. Flume runs were conducted for the conditions listed in table 1. All runs were based on a 3-ft diameter corrugated metal pipe with 3- by 1-inch corrugations. Two velocities (discharges) were tested at each embedment level and flow depth. Flow depths of 4.5, 6, and 9 inches represented 12, 16, and 25 percent of the culvert diameter, respectively. The flume runs are labeled as FpqrrVsDt where: 1) p represents the diameter of the culvert in feet, 2) q designates whether the run is a symmetrical half section (H) or full section (F), 3) rr represents the embedment level (in percent of the culvert diameter), 4) s represents velocity from lowest to highest in the matrix, and 5) t represents the depth from lowest to highest in the matrix. For example, F3H15V2D3 is a 3-ft diameter half-section run with the embedment at 15 percent of the diameter.

Table 1. Test matrix for the experimental flume.

Run ID	Flow Depth (inches)	Embedment (inches)	Flow Velocity (ft/s)
F3H00V1D1	4.5		
F3H00V1D2	6	0	0.71
F3H00V1D3	9		
F3H00V2D1	4.5		
F3H00V2D2	6	0	1.1
F3H00V2D3	9		
F3H15V1D1	4.5		
F3H15V1D2	6	5.4	0.71
F3H15V1D3	9		
F3H15V2D1	4.5		
F3H15V2D2	6	5.4	1.1
F3H15V2D3	9		
F3H30V1D1	4.5		
F3H30V1D2	6	10.8	0.71
F3H30V1D3	9		
F3H30V2D1	4.5		
F3H30V2D2	6	10.8	1.1
F3H30V2D3	9		

An embedment of zero inches represents the case with no embedment as shown in figure 8. Similarly, embedment representing 15 percent of the culvert diameter (5.4 inches) and 30 percent of the culvert diameter (10.8 inches) are illustrated in figure 9 and figure 10, respectively.

For the 15- and 30-percent embedment runs, the culvert bed material is coarse gravel with a mean particle diameter, D_{50}, equal to 0.472 inches. Bed roughness was achieved by gluing one layer of the gravel on the bottom of the flume. Flow depth is defined from the top of the bed.

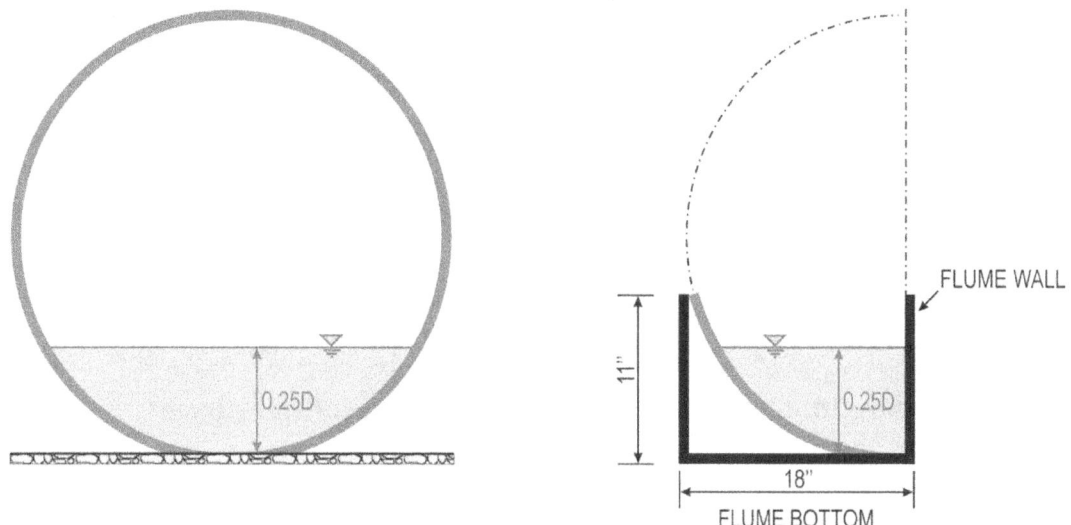

Figure 8. Illustration. Culvert with no embedment.

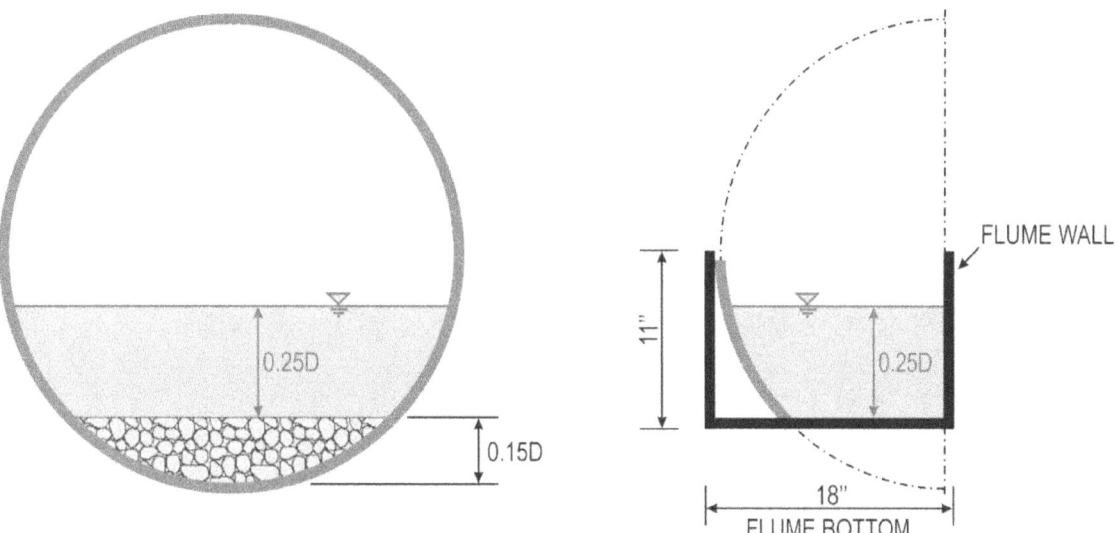

Figure 9. Illustration. Culvert with 15-percent embedment.

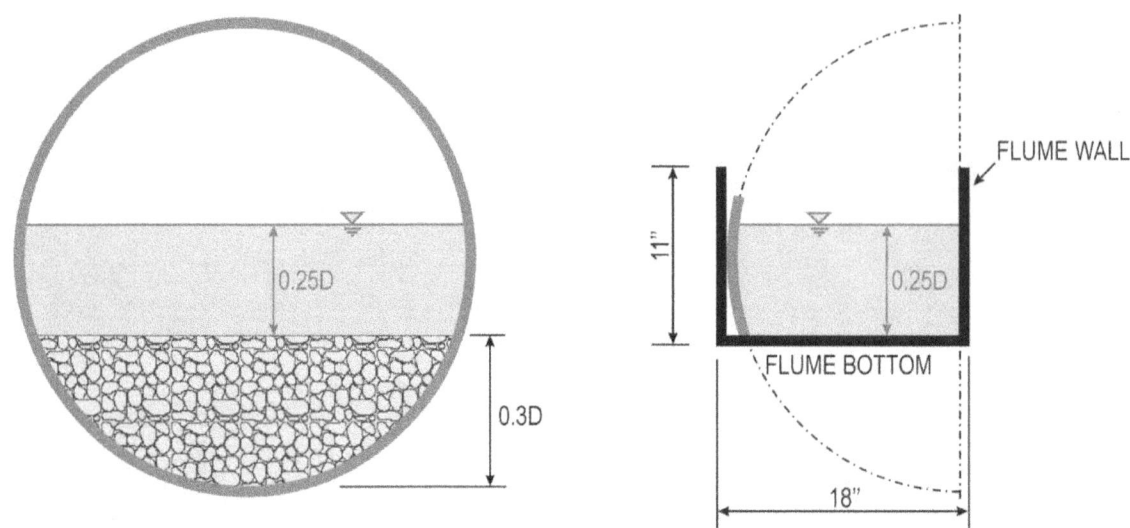

Figure 10. Illustration. Culvert with 30-percent embedment.

COMPARISON OF ADV AND PIV MEASUREMENTS

ADV and PIV results were compared to evaluate the robustness of the PIV results. Test runs used for this purpose had zero embedment and included three different flow depths and two different velocities. Figure 11 and figure 12 show a representative result for the case with a depth of 4.5 inches and velocity of 0.71 inches (run F3H00V1D1). The ADV results demonstrate that ADV is limited in its ability to collect data near fixed and water surface boundaries. Figure 13 and figure 14 summarize the flow fields for the case where depth is 9 inches and velocity is 1.1ft/s (run F3H00V2D3). ADV data in the area where the probe provides good readings are an important cross-reference for PIV data.

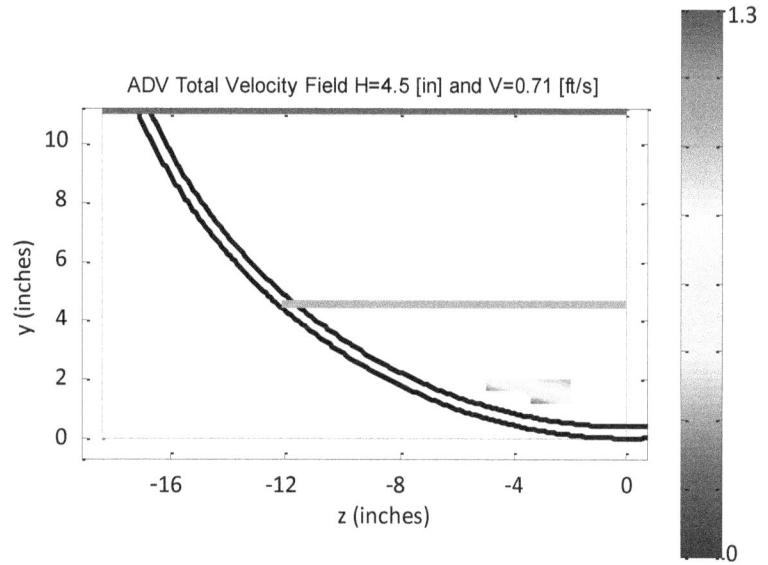

Figure 11. Graph. Run F3H00V1D1 velocity field using ADV.

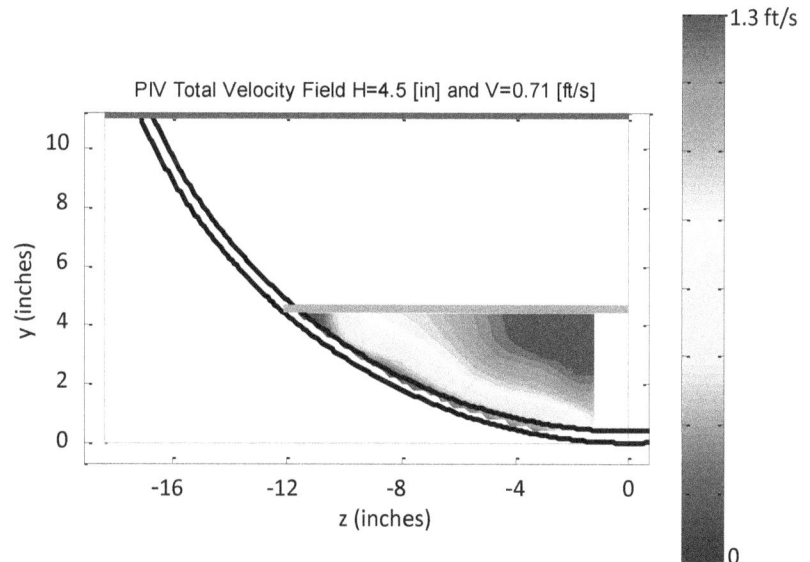

Figure 12. Graph. Run F3H00V1D1 velocity field using PIV.

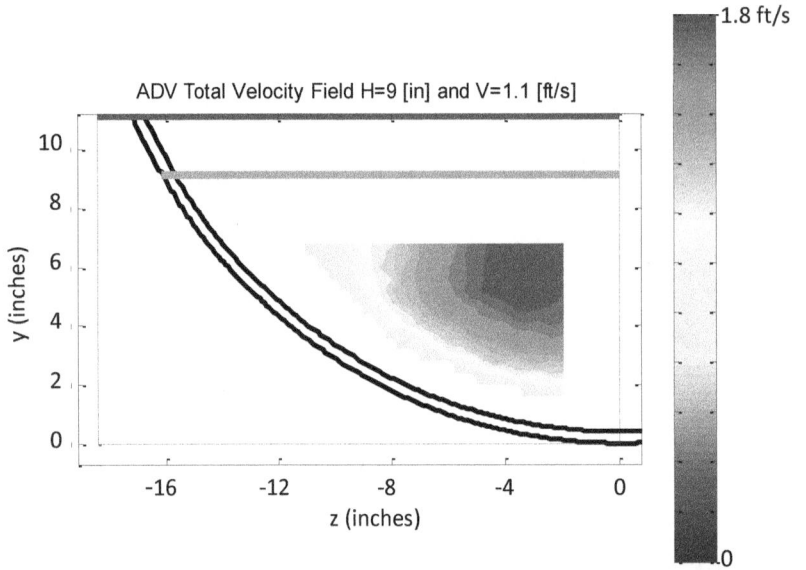

Figure 13. Graph. Run F3H00V2D3 velocity field using ADV.

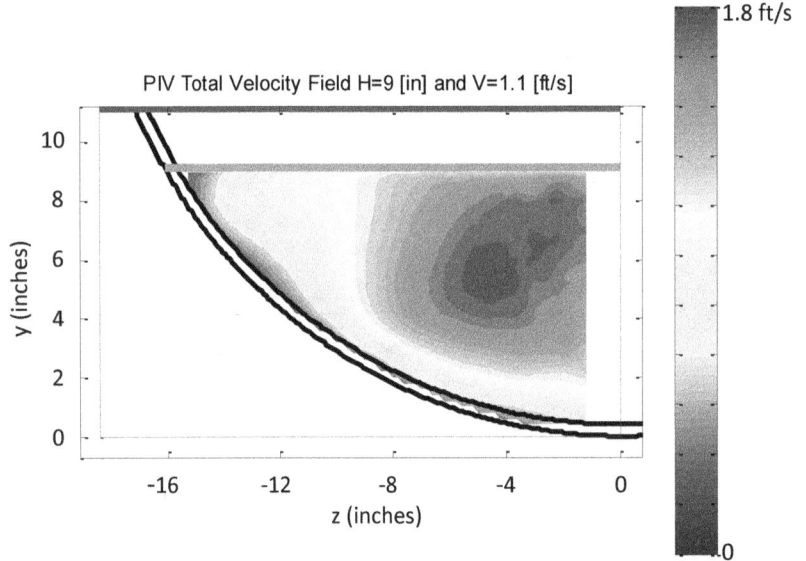

Figure 14. Graph. Run F3H00V2D3 velocity field using PIV.

CHAPTER 5. NUMERICAL MODELING

INTRODUCTION

Physical modeling requires substantial space and time resources that may limit the extent of inquiry. This research also used 3D CFD models to reproduce the complexity of the flows in culverts. CFD allowed for analysis of a more comprehensive set of conditions than was possible in the experimental flume, given the space and budget constraints of the study.

Several CFD modeling software packages are available for simulating flow conditions in a culvert. For this research, CD-adapco's STAR-CCM+ commercial software was used to analyze hydraulic conditions in several culvert configurations. The suitability of CFD modeling for fish passage analysis was assessed by comparing its results with experimental data obtained from the flume modeling.

The first CFD models were of the experimental flume. A 3D multiphase computer-aided design (CAD) model, as shown in figure 15, was created using Pro-ENGINEER. The CAD model consisted of three parts along the flow direction (x-axis): the intake, the barrel, and the diffuser. The culvert model considered in this phase of the study was a symmetrical half section of the culvert pipe having annular corrugations with no embedment.

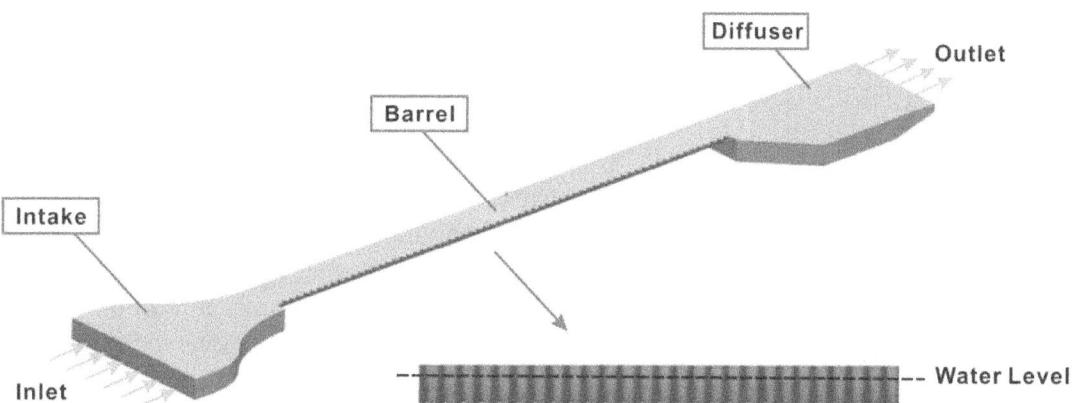

Figure 15. Illustration. 3D CAD model for initial CFD simulations.

CFD models are based on a computational mesh. The mesh must capture the boundary layer and preserve the near-wall thickness for the wall function to apply. Because the interface between water and air is an area with steep velocity gradients, mesh refinement volumetric control is used in this study. The resulting meshes are very detailed. For example, a model with a flow depth of 6 inches and an air domain of 3 inches may result in as many as 18 million mesh cells if the basic size is defined as 0.394 inches and the size of refined cells is 0.197 inches. The meshing model and the meshing refinement art are illustrated in figure 16.

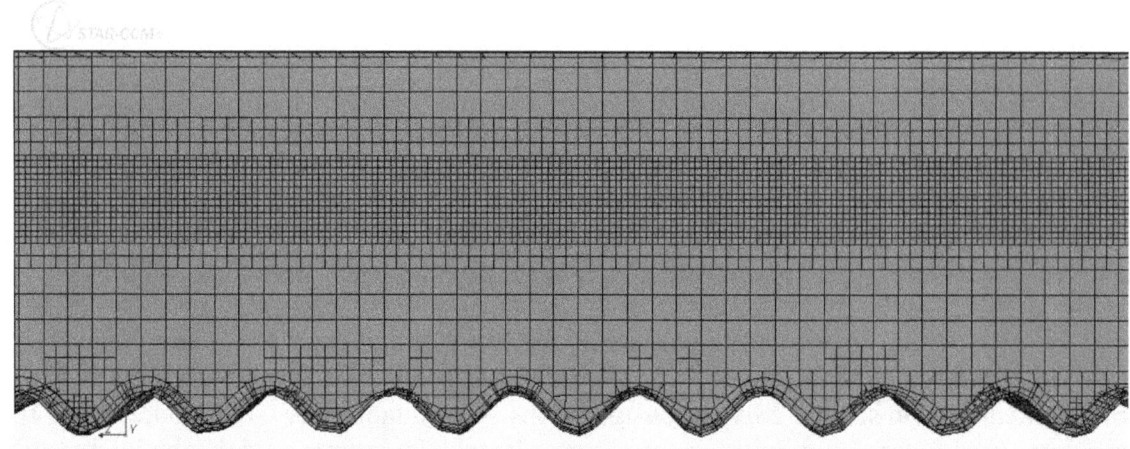

Figure 16. Illustration. Enlarged view of a mesh on the left hand side of the flow direction.

CFD MODELING APPROACH

A RANS numerical method has been employed in conjunction with a k-epsilon turbulence model and wall function technique for the simulation of complex 3D flow in the culvert. The volume of fluid (VOF) method, which captures the free surface profile through use of the variable known as the volume of fluid, was used in the two-phase (water and air) CFD model. The following physics models were enabled in the computational domain to replicate and simulate the realistic flow conditions: unsteady, liquid, segregated flow, constant density, k-ε turbulence, segregated fluid isothermal, and gravity.

Several strategies for CFD modeling were evaluated. Building a full two-phase (water and air) model of the entire culvert is the most flexible approach but is computationally intensive. One simplifying alternative is to build a single-phase (water only) model neglecting the role of the air layer. A second simplifying alternative is to truncate the model using only a fraction of the entire culvert length and applying a cyclic boundary condition. Both simplifications reduce run times.

Single Versus Two-Phase Modeling

The initial CFD modeling efforts explored representation of culvert flow conditions using the hypothesis of uniform flow in attempting to match conditions in the experimental flume. For uniform flow in a prismatic section, cross-sections exhibit the same velocity distribution throughout the model length. A two-phase (water and air) CFD model was prepared to simulate the change of the water surface within the culvert. A boundary condition is to define the water surface as the interface between the water and air layers. The VOF method was used in the STAR-CCM+ to handle the free water surface.[23]

The two-phase CFD model was created to strictly follow the physical dimensions of the half-section 3-ft diameter CMP culvert without embedment in the experimental flume. The simulations involved three water depths at the 0.71 ft/s flow velocity as described in table 1 for the flume. Specification of the CFD model included description of the flume tilt angle, flap gate opening angle, and roughness parameters. The flow conditions for these two-phase CFD model tests are listed in table 2.

Table 2. Flow conditions of multi-phase CFD model tests.

Model Characteristics	Model 1	Model 2	Model 3
Flow depth (inches)	4.5	6	9
Embedment (inches)	0	0	0
Air layer depth (inches)	2.5	3	2.5
Mean flow velocity (ft/s)	0.71	0.71	0.71
Flume tilt angle (degrees)	1	0.125	0.07
Flap gate opening angle with respect to the horizontal (degrees)	12.5	18	28

Three cross-sections were highlighted in the tests: 1) inlet of the barrel (section 1), 2) middle of the barrel (section 2), and 3) end of the barrel (section 3). Example results from the two-phase velocity distribution using model 2 (6-inch flow depth) are shown in figure 17. (Flow is from left to right.) The results show that at the intake and the diffuser, the velocity field is significantly different than the other two sections, but it is consistent from section 2 to section 3. In these cross-sections, the high-velocity zone is near the vertical smooth wall boundary and spans the air-water interface.

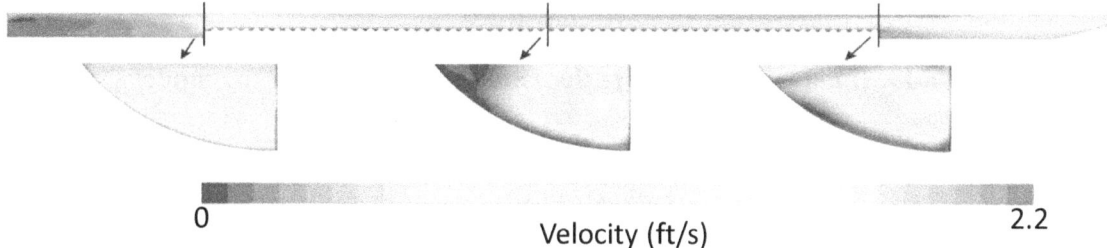

Figure 17. Illustration. Example velocity plots for the two-phase model.

Similarly, VOF distributions showing volume fraction of air (blue) and water (red) for the two-phase model are illustrated in figure 18. A nearly uniform height water–air interface exits along the length of the barrel. Taking the 0.5 VOF isoline as the air–water interface yields a water surface that coincides closely with the flume water level of 6 inches.

This result suggests that a single-phase model may be used to simulate the flow of water with reasonable accuracy. To further examine this possibility, the complete two-phase velocity distributions for section 2 (trough) is compared with the velocity distribution below the known flow depth. Figure 19 provides this comparison for the 3-inch flow depth of model 1. On the left is the complete distribution. On the right, only the velocity distribution below the known depth is plotted with the 0.5 VOF curve superimposed. As can be seen, this curve is close to the known depth. The results for 6- and 9-inch depths are illustrated in figure 20 and figure 21, respectively.

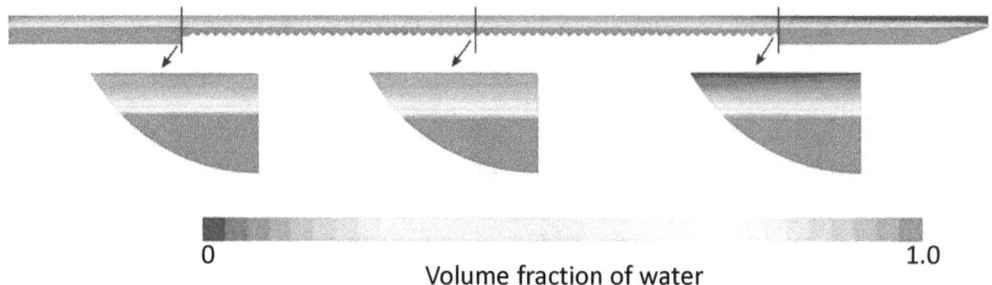

0   Volume fraction of water   1.0

Figure 18. Illustration. Example water fraction plots for the two-phase model.

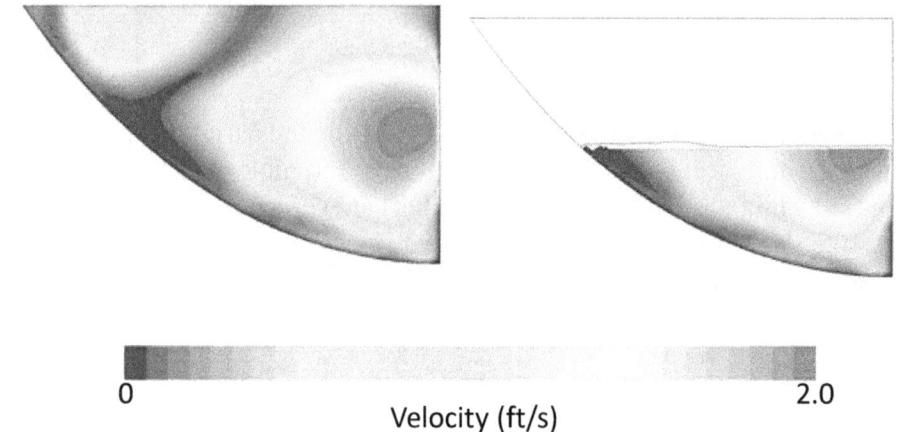

0   Velocity (ft/s)   2.0

Figure 19. Illustration. Velocity distribution comparison for model 1.

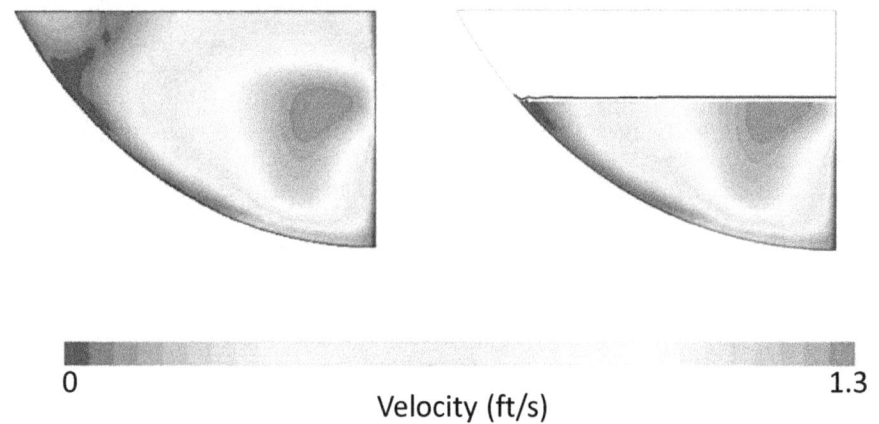

0   Velocity (ft/s)   1.3

Figure 20. Illustration. Velocity distribution comparison for model 2.

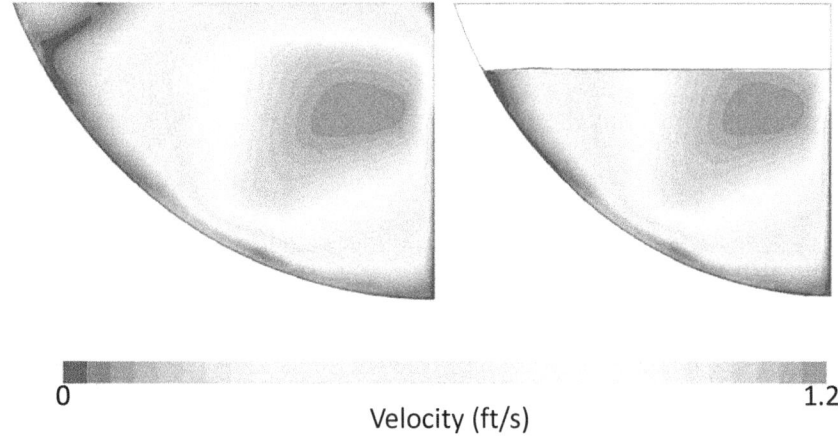

0 Velocity (ft/s) 1.2

Figure 21. Illustration. Velocity distribution comparison for model 3.

Building and running a two-phase full flume CFD model is time consuming. By trial and error, the appropriate flume tilt angle and flap gate opening angle are varied to achieve a uniform surface. When a nearly uniform water–air interface exists along the length of the barrel, as illustrated in figure 18, it appears possible that a single-phase model can be used to simulate the flow of water under the interface with reasonable accuracy. With this in mind, a single-phase model of the flow with a 6-inch water depth (model 2) was set up and executed. The velocity distribution results from the steady-state simulation are shown in figure 22. The same observations provided for the two-phase model shown in figure 17 are applicable for the single-phase model.

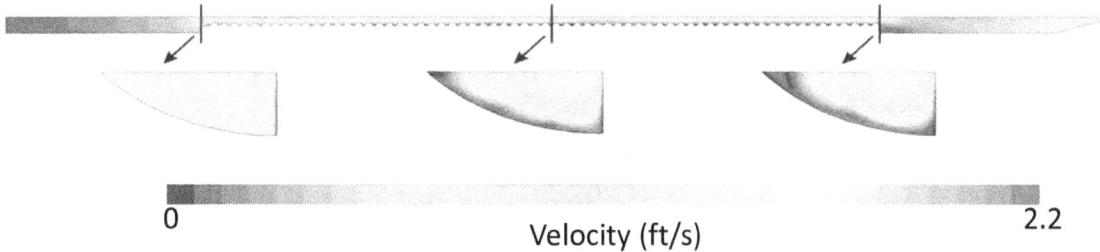

0 Velocity (ft/s) 2.2

Figure 22. Illustration. Velocity distribution for the single-phase model.

A comparison of the velocity distributions is provided in figure 23. The two-phase plots are cut at the 6-inch water depth, which is the same height as the single-phase flow domain. The 0.5-VOF curves are plotted on the top of the velocity contours and are slightly higher than the single-phase water flow depth. The velocity of the single-phase model is marginally larger than that of the two-phase model because in the multiphase flow model, the height of the free surface is computed, and it is slightly higher than the planned water depth. This slightly increases the cross-section area carrying water. For the purposes of defining velocities in a culvert that may inhibit fish passage for reasonably uniform flows, the small increase in water velocity when using the single-phase model is conservative. Because the single-phase flow CFD model greatly reduces computational resources and computation time, it is preferred.

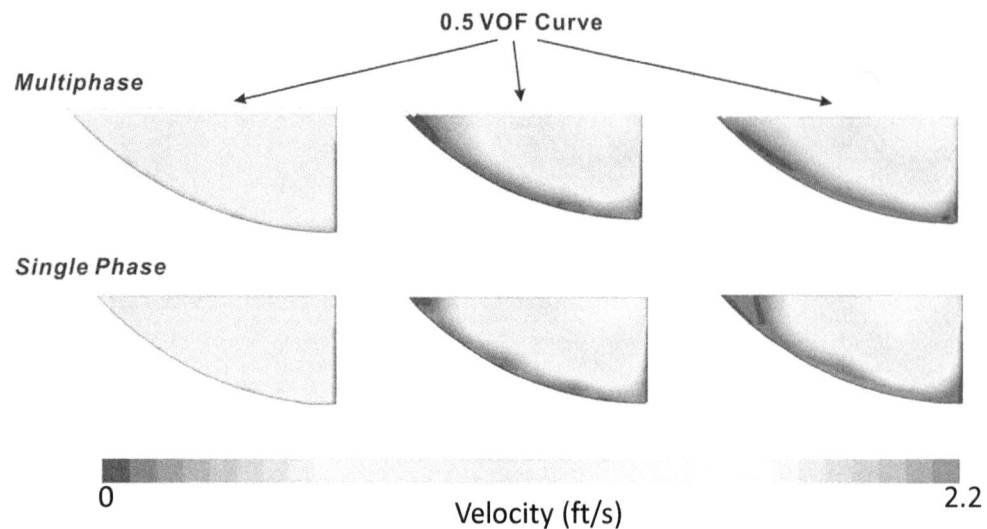

Figure 23. Illustration. Velocity comparison of the two-phase and single-phase models.

Truncated Single-Phase Modeling

The single-phase model significantly reduces the simulation time. Further reductions in run times were investigated using a single-phase model with cyclic boundary conditions at the section inlet and outlet. The run time reductions are achieved by modeling a truncated 3D section of the culvert (symmetric quarter of the culvert section with corrugations from trough to another trough) rather than the full culvert length as shown in figure 24.

Figure 24. Illustration. Truncated section of the cyclic single-phase model.

The cyclic boundary approach shortens the simulation time required to establish fully developed flow with a known mass flow rate. (With this approach, several test cases can be completed per day.) The periodic fully developed condition is achieved by creating a cyclic boundary condition where all outlet variables are mapped back to the inlet interface, except for the pressure because there is a pressure drop corresponding to the energy losses in the culvert section. The pressure jump needed to balance the pressure drop for the specified mass flow is iteratively computed by the CFD solver.

Variations in the resulting velocity distributions, and other flow parameters, of the flow field were examined by varying the base size of the mesh to obtain solutions that were effectively mesh independent.[24] Depending on the relative depth and the embedment, the number of mesh cells in the truncated model varied between 58,661 and 875,087. As previously described, more data points were taken near the boundary of the culvert than in the center of the cross-section to obtain more precise flow field data near the corrugated wall. With these observations, the truncated single-phase model with cyclic boundary conditions is used for subsequent CFD test runs. Test results indicated that for a single-phase model, the velocity simulation results of the cyclic boundary condition are similar to the non-cyclic boundary condition, which is illustrated in figure 25.

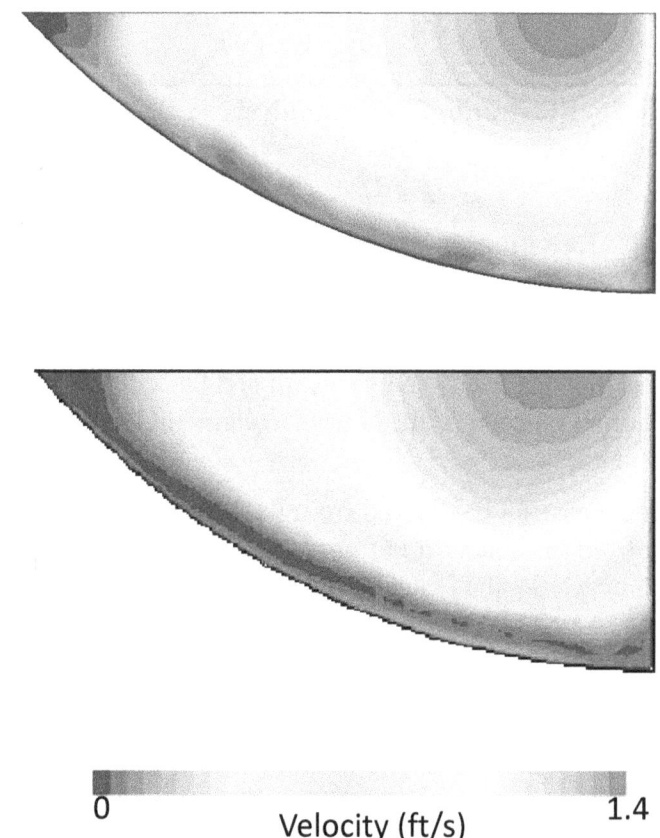

0   Velocity (ft/s)   1.4

Figure 25. Illustration. Velocity distributions for non-cyclic (upper) and truncated cyclic (lower) single-phase models.

MODELING THE BOUNDARY

To be effective, the CFD modeling must represent the roughness of the culvert walls as well as the bed material for the embedded culverts. This section describes this modeling effort as well as the methods for estimating composite roughness when more than one surface material influences the flow field.

Computing Manning's n from the CFD Analyses

The value of Manning's roughness is derived for the CFD runs using the force equilibrium equation based on figure 26 and the equation shown in figure 27. Taking a control volume of the fluid flow in an embedded pipe, the energy loss resulting from the shear stress on the wetted area (culvert walls and bed material) is reflected in a loss in pressure on the fluid flow face from the upstream end to the downstream end.

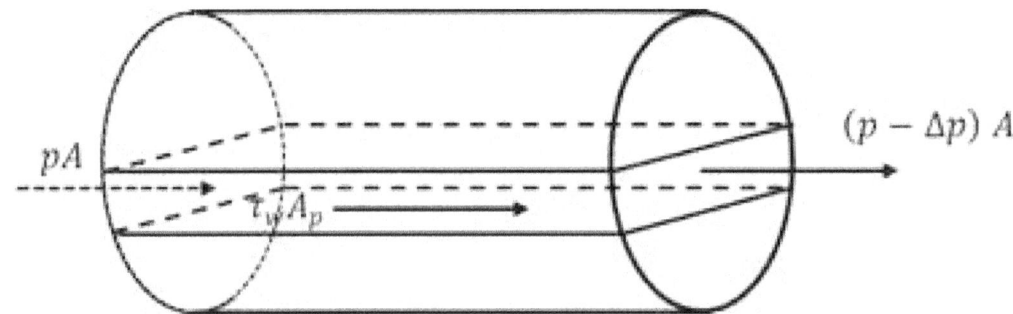

Figure 26. Illustration. Force analysis on the culvert pipe.

$$\tau_w A_p = \Delta p A$$

Figure 27. Equation. Force equilibrium on a control volume.

Where:
τ_w = Wetted perimeter shear stress, lb/ft^2.
A_p = Surface area of the wetted perimeter (culvert wall and bed), ft^2.
Δp = Change in pressure from one end of the control volume to the other, lb/ft^2.
A = Flow cross-sectional area, ft^2.

Because the pressures and areas can be derived from the CFD modeling, the wetted perimeter shear stress is computed from the equation in figure 27. Because the stresses on the wall and bed may differ with different materials, the result is the average shear stress across all the wetted surfaces.

The composite friction coefficient for a given flow condition may be estimated from the bed shear stress using the relation in figure 28.

$$\tau_w = \frac{f}{8}\rho V^2$$

Figure 28. Equation. Shear stress as a function of the friction factor.

Where:
f = Friction factor.
ρ = Water density.

The friction factor is related to the Manning's roughness coefficient by the equation in figure 29. The composite Manning's n value for each CFD run is determined by these relations.

$$n = \sqrt{\frac{f}{8g}} R_h^{1/6}$$

Figure 29. Equation. Relation of Manning's n to friction factor.

Culvert Wall

The annular corrugations of the metal culvert have been previously described. To validate that this representation provides the appropriate roughness characteristics for hydraulic modeling, several CFD runs were conducted with an 8-ft pipe flowing full. The Manning's n values from these runs were compared with those reported in HDS 5.[18] According to *Hydraulic Design of Highway Culverts* (HDS 5) (appendix B, figure B.3), the Manning's n for 6- by 2-inch annular corrugations on an 8-ft diameter CSP is 0.0344. This value is for high-flow design conditions, which are defined to be when $Q/D^{2.5}$ equals 4.0 $ft^{0.5}/s$.

The full-flow CFD runs using a 16-ft-long model resulted in a Manning's n value of 0.0325. This is less than 6 percent lower than the recommended value of HDS 5. Prior to running the 16-ft-long model, shorter models of 1 and 2 ft were also tested with CFD. The resulting Manning's n values were slightly lower than for the 16-ft model.

It should also be noted that the $Q/D^{2.5}$ value for the CFD is 0.29 $ft^{0.5}/s$, which is much lower than the values assumed for HDS 5. Although this indicates lower flow intensity, this difference is not expected to affect the computation of the Manning's n.

Bed Material

Characterizing bed roughness for natural rivers and streams within CFD models is challenging.[25] In the flume test runs, the culvert bed material was coarse gravel with a mean diameter, D_{50}, equal to 0.472 inches. For this size of material, the gravel bed boundary cannot simply be treated as a rough wall using wall functions because the centroid of the near wall computational cell must be at a position that is greater than the roughness height. For this size of gravel, the near wall mesh would be far too large for the analysis results to be mesh independent, as is required. The following three options to model flow parallel to a porous gravel bed were considered:[26]

- Treat the bed as porous media, with a flat interface dividing the two flow zones.
- Simulate the top layer of gravel with contours of the rough bed.
- Represent the rough bed as one or more layers of spheres.

Porous Media

A justification for simulating the roughness of a gravel bed using a porous media configuration is that in a gravel bed, there is shear induced flow within the bed. Depending upon the permeability of the bed material, various levels of sub-surface flow are possible. The porous media approach not only provides a method to simulate surface roughness, but it also includes the effects of the flow within the bed structure.

To model flow through a porous media, the *STAR-CCM+ User Guide* notes that it is not the details of the internal flow in the porous region that are of interest, but rather the macroscopic effect of the porous medium on the overall fluid flow.[27] The effect of the porous medium on the flow is defined using lumped parameters that are typically taken to be resistance coefficients for two sink terms in the momentum equation. One is linearly proportional to velocity and the other proportional to velocity squared. These parameters account for the effect of the distributed surface area of the porous media as a flow resistance on the bulk flow in the porous media and, consequently, affect the flow that borders the porous region. If the porous region represents a gravel bed in a culvert, then it should have the effect of a very rough wall with a small slip condition at the interface that is determined by the flow resistance of the porous media.

Various CFD tests were conducted to determine whether a porous media model appears to be a viable way to model large-diameter gravel in the bed of a culvert. Some of the results obtained appeared reasonable, while others appeared to be physically unrealistic. The physically unrealistic results did not appear to capture the viscous transport of momentum from the open-flow region into the porous-flow region. This result was most likely a consequence of mesh refinement in the vicinity of the porous media interface. Because a boundary layer with a small slip condition does exist at the porous media interface, and wall functions for this model are not available to determine shear stress at the interface, using a sufficiently refined grid to resolve the velocity profile near the interface is important.

An additional difficulty with using the porous media approach was that comparisons would be made to physical modeling results and not field measurements. Because the physical modeling of the bed material was achieved by gluing one layer of the gravel on the bottom of the flume, there would be no flow through the bed material, leaving the porous media approach unsuitable for this project.

Rough Bed Contouring

Another method was to model the gravel roughness by simulating the top layer of the gravel by detailed contouring. The CAD model for the contoured bed was created in Pro-ENGINEER. The challenge in selecting the appropriate contouring to mimic a gravel bed is the potential for introducing numerical error after importing the CAD bed model into STAR-CCM+. Because numerical errors may be caused by sharp angles between grids, two quarter-circular arcs with the appropriate diameter are used to create smooth connections. Figure 30 illustrates the method of creating a D_{50} equal to 0.944 inches. Figure 31 shows a part of the model with the contours simulating a gravel bed. This approach for representing the gravel bed was used for the embedded culvert runs listed in table 4 through table 7.

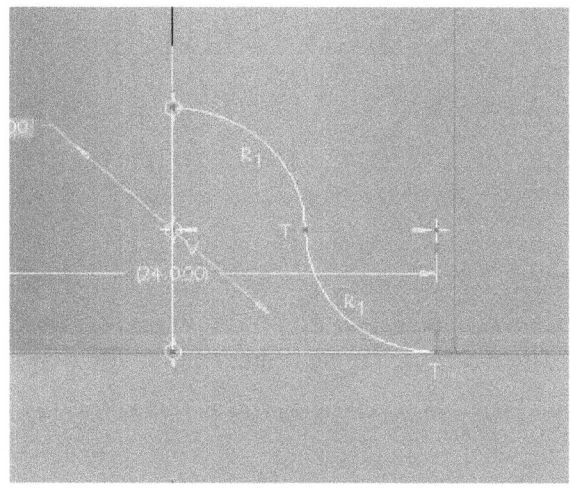

Figure 30. Illustration. Gravel contour method in the CAD model.

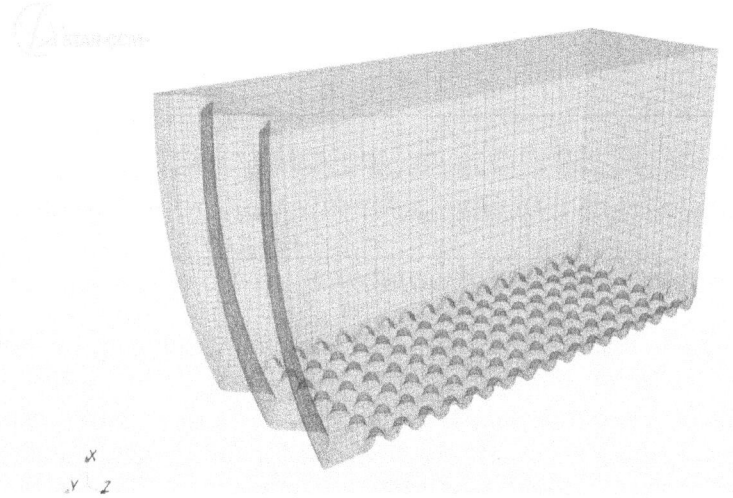

Figure 31. Illustration. Gravel bed surface created by contouring.

Spheres

The final method was to model the gravel roughness as one or more layers of uniform spheres. The CAD model for this approach was created in SolidWorks. Spheres with diameters (D_{50} values) of 0.39, 0.87, 1.34, and 2.28 inches were used in CFD runs primarily to refine the modeling estimates of Manning's n and for estimating composite roughness values. It was found that when the particle size is small, the porosity of the bed material is not properly simulated with a single layer of particles, as shown in figure 32. This may be because the boundary below is a smooth, solid surface and does not represent a gravel bed well. To address this problem, a second layer of particles, as shown in figure 33, was added and the results compared with the single layer runs. A single layer worked well for the two larger diameter spheres, but a double layer was needed for the 0.39 and 0.87 inch D_{50} runs.

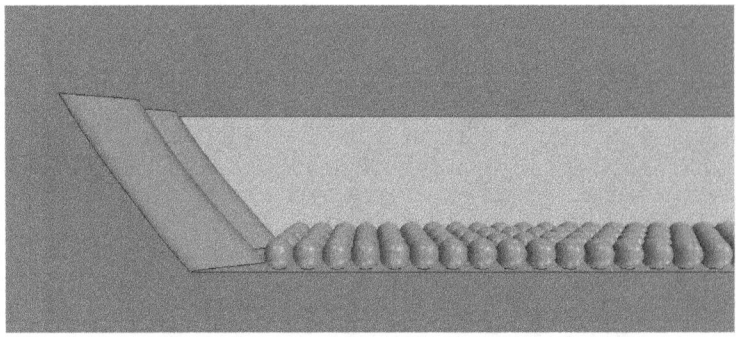
Figure 32. Illustration. Single sphere layer for bed roughness.

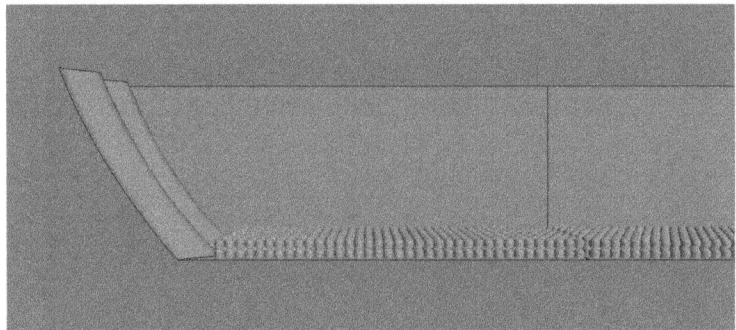
Figure 33. Illustration. Double sphere layer for bed roughness.

Comparison With Empirical Estimates of Bed Roughness

Results from the CFD modeling may be compared with Manning's roughness estimates using empirical relations such as the Blodgett equation for bed roughness shown in figure 34.[1] Other relations are available and discussed in *Aquatic Organism Passage Design Guidelines for Culverts* (HEC 26), but the Blodgett equation performs well for the range of conditions considered in this study.[1]

$$n = \frac{\alpha\, y^{1/6}}{2.25 + 5.23\, log\left(\dfrac{y}{D_{50}}\right)}$$

Figure 34. Equation. Blodgett's equation for bed roughness.

Where:
$\alpha = 0.262$ for customary units and 0.319 for metric units.

Because the CFD roughness values are composite values (bed and culvert wall), the bed material roughness was not directly estimated in the CFD runs and could not be compared with the Blodgett estimates. The ability of CFD to represent the bed roughness is assessed indirectly through comparison of overall composite roughness.

Further complicating the comparison is that the CFD representation of the bed material using the spheres includes particles of only one size. The Blodgett equation uses the D_{50} as a defining characteristic of the bed material but anticipates that the bed material has particles both smaller and larger than this value.

Composite Roughness

To fully evaluate the CFD modeling of roughness, the composite roughness of an embedded culvert must consider the combined effect of the bed material and the culvert material. The CFD computation method presented previously directly produces a composite roughness because the combined shear force from the bed and culvert wall is jointly considered. These CFD estimates of the composite Manning's n are compared with composite Manning's n estimates from each of three compositing methods that use a separate value of roughness for the culvert wall and the Blodgett estimate of bed material roughness.

Compositing Methods

A detailed description of compositing methods is provided in appendix C of HEC 26.[1] It includes three alternative methods. One approach is a simple weighting of n values based on wetted perimeter. This approach is used by the software tool FishXing and is described by the equation in figure 35.

$$n_c = \left[\frac{P_b n_b + P_w n_w}{P_b + P_w} \right]$$

Figure 35. Equation. Linear compositing.

Where:
n_c = Composited n-value for the culvert.
P_b = Wetted perimeter of the bed material in the culvert.
n_b = n-value of the bed material in the culvert.
P_w = Wetted perimeter of the culvert walls above the bed material.
n_w = n-value of the culvert material.

A second approach is applied by the software tools HY-8 and Hydrologic Engineering Centers River Analysis System (HEC-RAS), described in HDS 5, and recommended by Tullis.[28, 18, 29] It uses an exponent other than unity in weighting the roughness values. This approach is described by the equation in figure 36.

$$n_c = \left[\frac{P_b n_b^{1.5} + P_w n_w^{1.5}}{P_b + P_w} \right]^{2/3}$$

Figure 36. Equation. Exponential compositing.

A third approach considers that the energy lost to friction along the wetted perimeter is a function of the roughness and the boundary force applied to the boundary. This approach recognizes that for open channel flow, the hydrostatic forces on the culvert side wall are not equivalent to those on the bed. The approach incorporates this concept by introducing a wall shape coefficient in the weighting that is fully described in HEC 26.[1] The equation is presented in figure 37.

$$n_c = \left[\frac{P_b n_b + c_y P_w n_w}{P_b + c_y P_w} \right]$$

Figure 37. Equation. Linear compositing with wall shape coefficient.

Where:
c_y = Wall shape coefficient, between 0 and 1.

The wall shape coefficient is 0.5 for a box culvert. Because the shallow flows in embedded culverts approximate a rectangular shape for the flow field, the 0.5 value for wall shape coefficient is appropriate for these conditions.

Analysis and Recommendations

A total of 18 CFD runs were completed to compare estimates of composite Manning's roughness. These included three flow depths, three particle sizes, and two levels of embedment within an 8-ft diameter CSP culvert. The bed material was represented using the spherical approach described previously with sphere diameters of 0.39, 0.87, and 1.34 inches. All runs used the same average flow velocity. From each of these runs, a composite Manning's n was developed using the previously described methodology.

These CFD-derived values are compared with values derived from the HDS 5 value of 0.0344 for the culvert wall and Blodgett's equation for the bed material given the D_{50} and flow depth for each case. The individual material values were then composited using each of the three methodologies outlined. The root mean square (RMS) difference between the CFD and analytically derived values was computed.

The RMS difference ranged from 0.0040 for the linear compositing method with the wall shape coefficient to 0.0047 for the exponential compositing method. These differences range from 12 to 14 percent of the HDS 5 Manning's n value of 0.0344 for the culvert wall. Figure 38 shows a plot of the CFD values versus the HDS 5/Blodgett values using the exponential compositing method. Although the data are scattered, the CFD values tend to underestimate the analytically derived estimates.

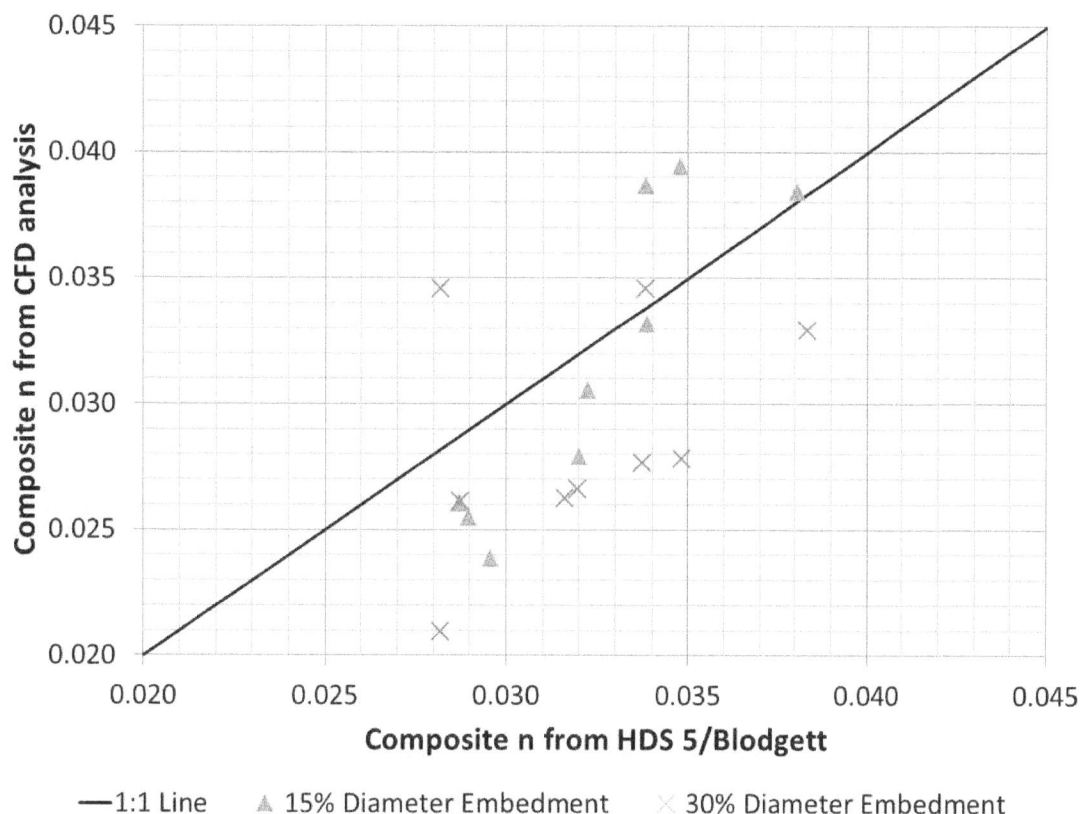

Figure 38. Graph. Comparison of composite Manning's n.

Several factors may contribute to the differences in composite Manning's n computed from the CFD modeling and the analytical methods. These include the compositing method, the use of the HDS 5 culvert value, the accuracy of the Blodgett equation for bed material roughness, and the CFD surface modeling approach. Because the latter has already been addressed, variations in the other three factors are evaluated.

Alternative values of the culvert wall and bed material roughness were selected, and the resulting composite Manning's n from each of the three compositing methods was computed. Table 3 summarizes the RMS differences for these analytical scenarios compared with the CFD results.

The first row in the table summarizes the results when using the HDS 5 culvert value and Blodgett estimates without adjustment. This is the scenario discussed previously. In the second scenario summarized in the table, the HDS 5 culvert value of 0.0344 is replaced with the Manning's n from the CFD modeling of the culvert flowing full without any embedment (0.0325). The third scenario retains the 0.0325 value for the culvert wall and employs a multiplier on the Blodgett estimates for the bed material roughness as shown in figure 39. The multiplier was selected to minimize the RMS difference. The Blodgett multiplier is intended to compensate for any systematic under or over estimate of the bed material Manning's n. The resulting multiplier was 0.93, meaning that when all Blodgett estimates are reduced by 7 percent, the smallest RMS difference is obtained.

$$n = k \left(\frac{\alpha \, y^{1/6}}{2.25 + 5.23 \, log\left(\frac{y}{D_{50}}\right)} \right)$$

Figure 39. Equation. Blodgett's equation for bed roughness with multiplier.

Where:

k = Blodgett multiplier.

Table 3. Summary of composite Manning's n analyses.

Blodgett Multiplier, k	Culvert Wall Manning's n	RMS Difference: Linear	RMS Difference: Exponential	RMS Difference: Linear with Wall Coefficient
1.0	0.0344	0.0046	0.0047	0.0040
1.0	0.0325	0.0044	0.0044	0.0042
0.93	0.0325	0.0041	0.0041	0.0040

Among the scenarios analyzed and compositing methods considered, there are no clear conclusions about the superiority of one approach over another. Although the RMS differences are lower with the wall coefficient method in all three scenarios, the differences between the three methods are not significant. Therefore, the exponential method is recommended because it already enjoys widespread acceptance.

TEST MATRICES

Test matrices were developed to achieve several project objectives. All CFD runs are labeled as CpqrrVsDt where: 1) p represents the diameter of the culvert in feet, 2) q designates whether the run is a symmetrical half section (H) or full section (F), 3) rr represents the embedment level (in percent of the culvert diameter), 4) s represents a velocity index from lowest to highest in the matrix, and 5) t represents the depth index from lowest to highest in the matrix. For example, C6F15V2D3 is a 6-ft-diameter full-section run with the embedment at 15 percent of the diameter. The depth and velocity for the run are the second and third values, respectively, from the matrix.

The objective of matrix CFD1 (see table 4) is to validate CFD results by comparing the modeled velocity fields with those measured in the experimental flume. All runs are based on a symmetrical half-section 3-ft corrugated metal pipe with 3-inch by 1-inch corrugations. Two velocities (discharges) were tested at each embedment and flow depth. Flow depths of 4.5, 6, and 9 inches represented 12, 16, and 25 percent of the culvert diameter, respectively. Embedment of 0 inches represents the case with no embedment, while embedment depths of 5.4 inches and 10.8 inches represent 15 and 30 percent of the culvert diameter, respectively. In the 15- and 30-percent embedment runs, a bed material with a D_{50} of 0.472 inches was simulated using the bed contouring approach.

Table 4. Matrix CFD1 for 3-ft CMP (half section).

Run ID	Flow Depth (inches)	Embedment (inches)	Flow Velocity (ft/s)
C3H00V1D1	4.5		
C3H00V1D2	6	0	0.71
C3H00V1D3	9		
C3H00V2D1	4.5		
C3H00V2D2	6	0	1.1
C3H00V2D3	9		
C3H15V1D1	4.5		
C3H15V1D2	6	5.4	0.71
C3H15V1D3	9		
C3H15V2D1	4.5		
C3H15V2D2	6	5.4	1.1
C3H15V2D3	9		
C3H30V1D1	4.5		
C3H30V1D2	6	10.8	0.71
C3H30V1D3	9		
C3H30V2D1	4.5		
C3H30V2D2	6	10.8	1.1
C3H30V2D3	9		

The objectives of matrix CFD2 (see table 5) are to validate CFD results by comparing the half-section computations with full-section computations. All runs are based on a 3-ft corrugated metal pipe with 3-inch by 1-inch corrugations. Two velocities (discharges) were tested at each embedment and flow depth. Flow depths of 4.5, 5, 6, 7, 8, 9, and 10 inches were tested. Embedment of 0 inches represents the case with no embedment, while embedment depths of 5.4 inches and 10.8 inches represent 15 and 30 percent of the culvert diameter, respectively. For the 15 and 30 percent embedment runs, a bed material with a D_{50} of 0.472 inches was simulated using the bed contouring approach.

Table 5. Matrix CFD2 for 3-ft CMP (full section).

Run ID	Flow Depth (inches)	Embedment (inches)	Flow Velocity (ft/s)
C3F00V1D1	4.5		
C3F00V1D2	5		
C3F00V1D3	6		
C3F00V1D4	7	0	0.71
C3F00V1D5	8		
C3F00V1D6	9		
C3F00V1D7	10		
C3F00V2D1	4.5		
C3F00V2D2	5		
C3F00V2D3	6		
C3F00V2D4	7	0	1.1
C3F00V2D5	8		
C3F00V2D6	9		
C3F00V2D7	10		
C3F15V1D1	4.5		
C3F15V1D2	5		
C3F15V1D3	6		
C3F15V1D4	7	5.4	0.71
C3F15V1D5	8		
C3F15V1D6	9		
C3F15V1D7	10		
C3F15V2D1	4.5		
C3F15V2D2	5		
C3F15V2D3	6		
C3F15V2D4	7	5.4	1.1
C3F15V2D5	8		
C3F15V2D6	9		
C3F15V2D7	10		
C3F30V1D1	4.5		
C3F30V1D2	5		
C3F30V1D3	6		
C3F30V1D4	7	10.8	0.71
C3F30V1D5	8		
C3F30V1D6	9		
C3F30V1D7	10		

C3F30V2D1	4.5		
C3F30V2D2	5		
C3F30V2D3	6		
C3F30V2D4	7	10.8	1.1
C3F30V2D5	8		
C3F30V2D6	9		
C3F30V2D7	10		

The objective of matrix CFD3 (see table 6) is to consider scaled roughness and slope. All runs are based on a 6-ft corrugated metal pipe with 6-inch by 2-inch corrugations. Two velocities (discharges) were tested at each embedment and flow depth. Flow depths of 6, 12, and 18 inches represented 8, 16, and 25 percent of the culvert diameter, respectively. All runs are models of full culvert sections. Embedment of zero inches represents the case with no embedment, while embedment depths of 10.8 inches and 21.6 inches represent 15 and 30 percent of the culvert diameter, respectively. In the 15- and 30-percent embedment runs, a bed material with a D_{50} of 0.944 inches was simulated using the bed contouring approach. Roughness for the CMP and embedment material is based on detailed representation of the CMP and the contoured bed material. The 6-ft culvert runs are scaled from the 3-ft runs based on the Froude number.

Table 6. Matrix CFD3 for 6-ft CMP.

Run ID	Flow Depth (inches)	Embedment (inches)	Flow Velocity (ft/s)
C6F00V1D1	6		
C6F00V1D2	12	0	1.0
C6F00V1D3	18		
C6F00V2D1	6		
C6F00V2D2	12	0	1.56
C6F00V2D3	18		
C6F15V1D1	6		
C6F15V1D2	12	10.8	1.0
C6F15V1D3	18		
C6F15V2D1	6		
C6F15V2D2	12	10.8	1.56
C6F15V2D3	18		
C6F30V1D1	6		
C6F30V1D2	12	21.6	1.0
C6F30V1D3	18		
C6F30V2D1	6		
C6F30V2D2	12	21.6	1.56
C6F30V2D3	18		

The objective of matrix CFD4 (see table 7) is to investigate velocity distributions at a large (prototype) scale. All runs are based on an 8-ft corrugated metal pipe with 6- by 2-inch corrugations. Two velocities (discharges) were tested at each embedment and flow depth. Flow depths of 7.7, 15.4, and 24 inches represented 8, 16, and 25 percent of the culvert diameter, respectively. The velocities were chosen to represent a realistic range of velocities of concern for fish passage at this scale. All runs are models of full culvert sections. Embedment of 0 inches represents the case with no embedment, while embedment depths of 14.4 and 28.8 inches represent 15 and 30 percent of the culvert diameter, respectively. For the 15- and 30-percent embedment runs, a bed material with a D_{50} of 0.944 inches was simulated using the bed contouring approach. The 8-ft culvert runs are not scaled from the 3-ft runs.

Table 7. Matrix CFD4 for 8-ft CMP.

Run ID	Flow Depth (inches)	Embedment (inches)	Flow Velocity (ft/s)
C8F00V1D1	7.7		
C8F00V1D2	15.4	0.0	1.0
C8F00V1D3	24.0		
C8F00V2D1	7.7		
C8F00V2D2	15.4	0.0	3.0
C8F00V2D3	24.0		
C8F15V1D1	7.7		
C8F15V1D2	15.4	14.4	1.0
C8F15V1D3	24.0		
C8F15V2D1	7.7		
C8F15V2D2	15.4	14.4	3.0
C8F15V2D3	24.0		
C8F30V1D1	7.7		
C8F30V1D2	15.4	28.8	1.0
C8F30V1D3	24.0		
C8F30V2D1	7.7		
C8F30V2D2	15.4	28.8	3.0
C8F30V2D3	24.0		

COMPARISON OF PIV AND CFD

The data from physical modeling provided a reliable means to calibrate and validate the CFD modeling. For each flow condition specified in the test matrix for physical modeling (see table 1), comparisons were made with velocity data from CFD modeling (matrix CFD1). The physical modeling included runs with 3 water depths, 2 flow velocities, and 3 embedment levels to produce 18 data sets.

Section Comparison

Velocities measured in the flume by PIV and estimated by the CFD model were compared using the root mean square error (RMSE) of data points measured/computed in the cross-section. (The PIV measurements from the physical model are considered to be the "correct" values.) Interpolating both data sources to a 0.197- by 0.197-inch grid, the RMSE is defined in the equation in figure 40.

$$RMSE = \sqrt{\frac{\sum (V_{i,1} - V_{i,2})^2}{n}}$$

Figure 40. Equation. RMSE.

Where:
$RMSE$ = Root mean square error in velocity, ft/s.
$V_{i,1}$ = Velocity at the ith point in the cross-section as computed by CFD, ft/s.
$V_{i,2}$ = Velocity at the ith point in the cross-section as measured by PIV, ft/s.
n = Number of velocity measurements available in the cross-section.

A relative percentage error is defined in the equation in figure 41. The relative errors for each run are listed in table 8 and range from 10.3 to 24.0 percent.

$$RE = \frac{RMSE}{V_a}$$

Figure 41. Equation. Relative error.

Where:
RE = Relative error, dimensionless.
V_a = Average velocity (nominal) in the cross-section, ft/s.

Color-coded velocity plots provide visual evidence of the agreement between CFD simulations and the physical experiments. The condition with zero embedment with a flow depth of 4.5 inches and a velocity of 0.71 ft/s (run F3H00V1D1) is shown in figure 42. Similarly, the condition with a depth of 9 inches and a velocity of 1.1 ft/s (run F3H00V2D3) is shown in figure 43. Both show good agreement.

Additional runs were completed with embedment at 15 and 30 percent of the culvert diameter. Results for the 30-percent embedment level with a flow depth of 4.5 inches and a velocity of 0.71 ft/s (run F3H30V1D1) are shown in figure 44. Similarly, the condition with a depth of 9 inches and a velocity of 1.1 ft/s (run F3H30V2D3) is shown in figure 45. Both show good agreement.

The gridded point velocities were averaged to determine the average cross-sectional velocity for each run as estimated in the flume by PIV and in the CFD modeling. These are compared in figure 46 and show reasonable agreement. This comparison supports the adequacy of CFD modeling for the purposes of this project.

ADV measurements, limited by the relatively small measurable area in the culvert cross-section (especially in shallow conditions), did not compare very well with the average velocity from CFD results. These limits were discussed previously.

Table 8. Relative error for cross-section velocity measurements.

Flume Run ID	CFD Run ID	Embedment (inches)	Velocity (ft/s)	Water depth (inches)	Relative Error (percent)
F3H00V1D1	C3H00V1D1	0	0.71	4.5	24.0
F3H00V1D2	C3H00V1D2	0	0.71	6	20.0
F3H00V1D3	C3H00V1D3	0	0.71	9	20.0
F3H00V2D1	C3H00V2D1	0	1.1	4.5	28.4
F3H00V2D2	C3H00V2D2	0	1.1	6	18.6
F3H00V2D3	C3H00V2D3	0	1.1	9	21.8
F3H15V1D1	C3H15V1D1	5.4	0.71	4.5	12.3
F3H15V1D2	C3H15V1D2	5.4	0.71	6	20.9
F3H15V1D3	C3H15V1D3	5.4	0.71	9	10.6
F3H15V2D1	C3H15V2D1	5.4	1.1	4.5	10.3
F3H15V2D2	C3H15V2D2	5.4	1.1	6	14.5
F3H15V2D3	C3H15V2D3	5.4	1.1	9	10.6
F3H30V1D1	C3H30V1D1	10.8	0.71	4.5	20.3
F3H30V1D2	C3H30V1D2	10.8	0.71	6	18.2
F3H30V1D3	C3H30V1D3	10.8	0.71	9	12.8
F3H30V2D1	C3H30V2D1	10.8	1.1	4.5	19.9
F3H30V2D2	C3H30V2D2	10.8	1.1	6	16.4
F3H30V2D3	C3H30V2D3	10.8	1.1	9	11.2

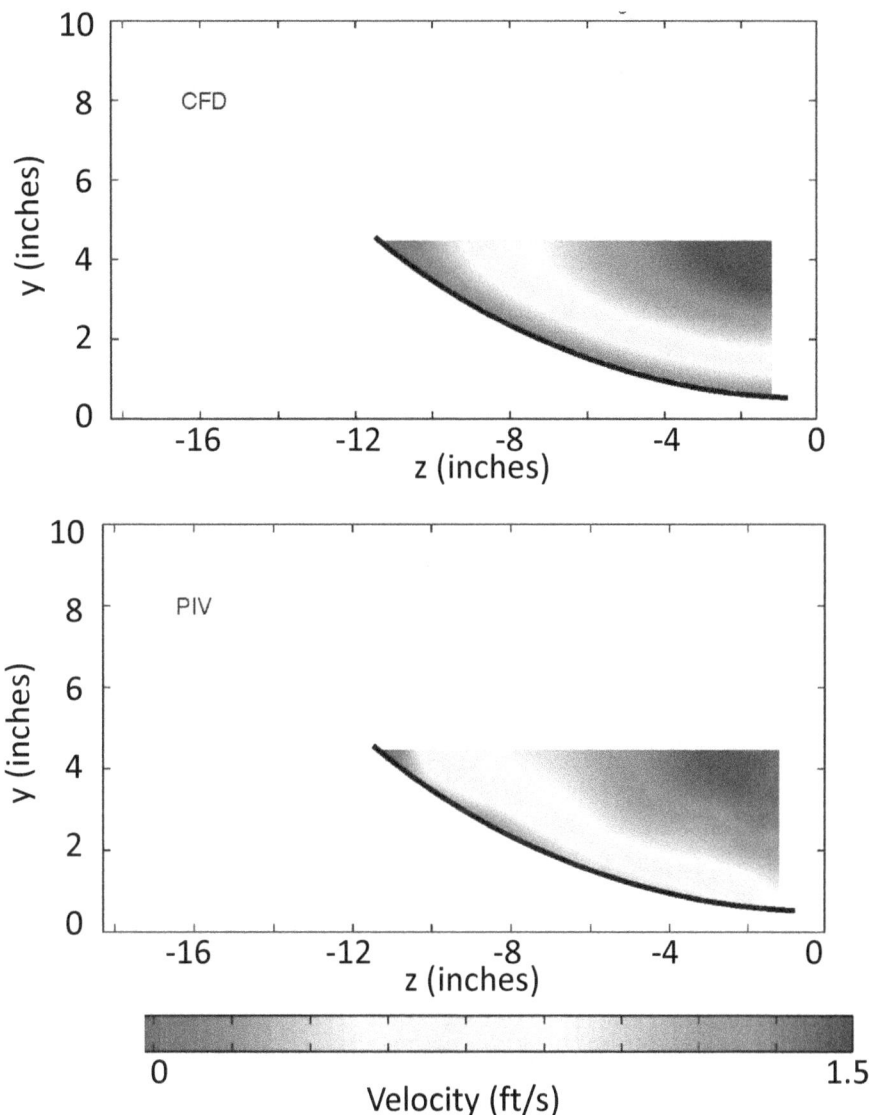

Figure 42. Illustration. Comparison of CFD and PIV results for run F3H00V1D1.

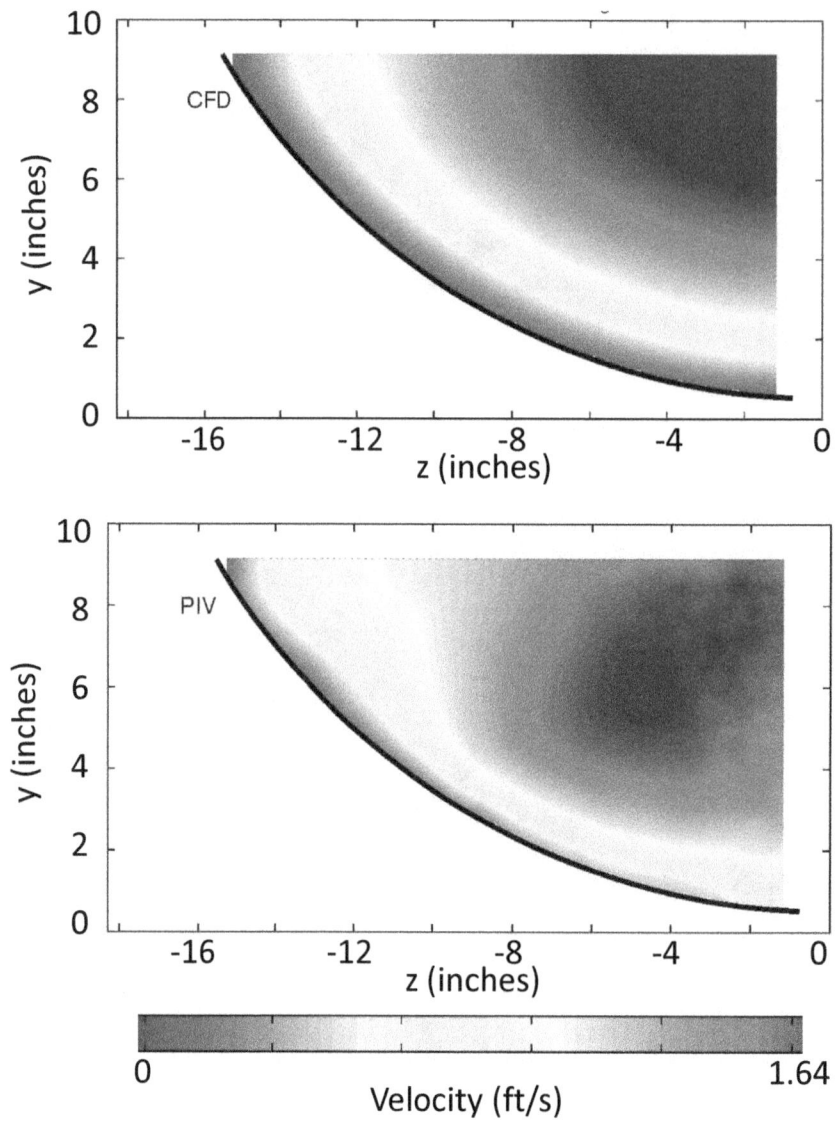

Figure 43. Illustration. Comparison of CFD and PIV results for run F3H00V2D3.

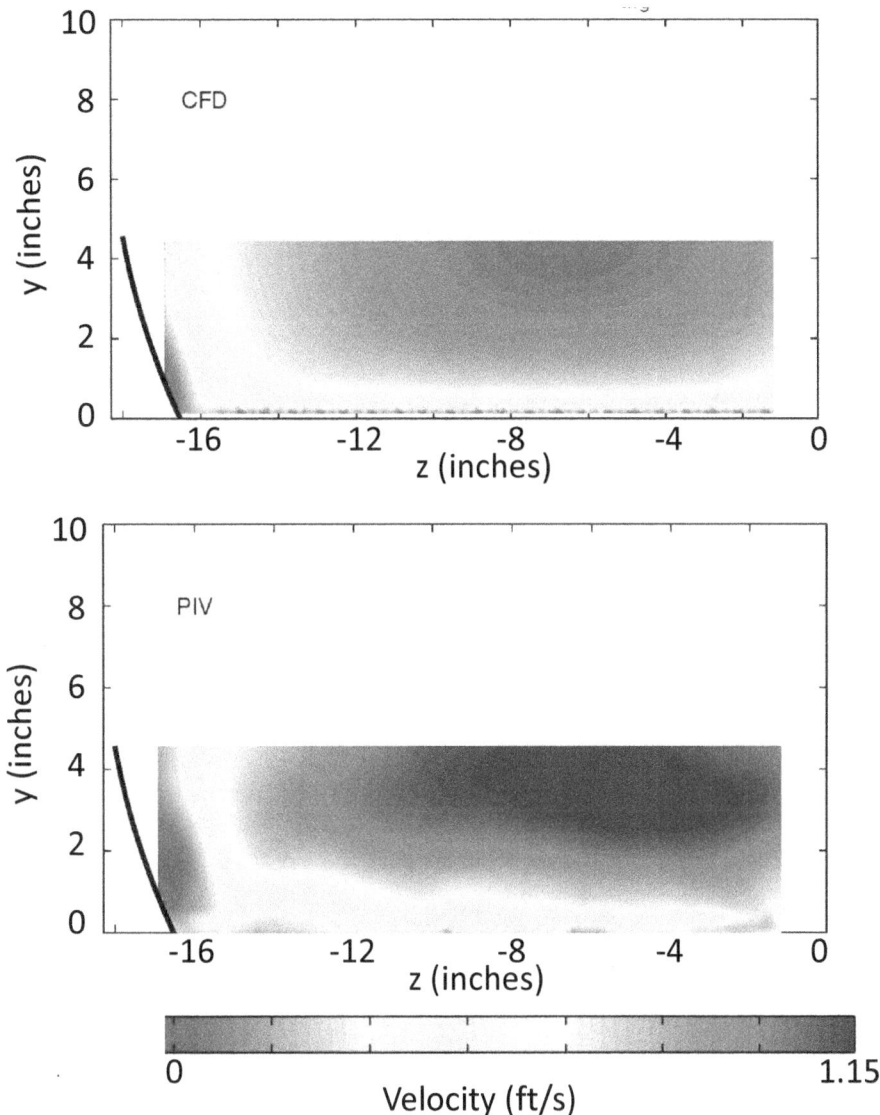

Figure 44. Illustration. Comparison of CFD and PIV results for run F3H30V1D1.

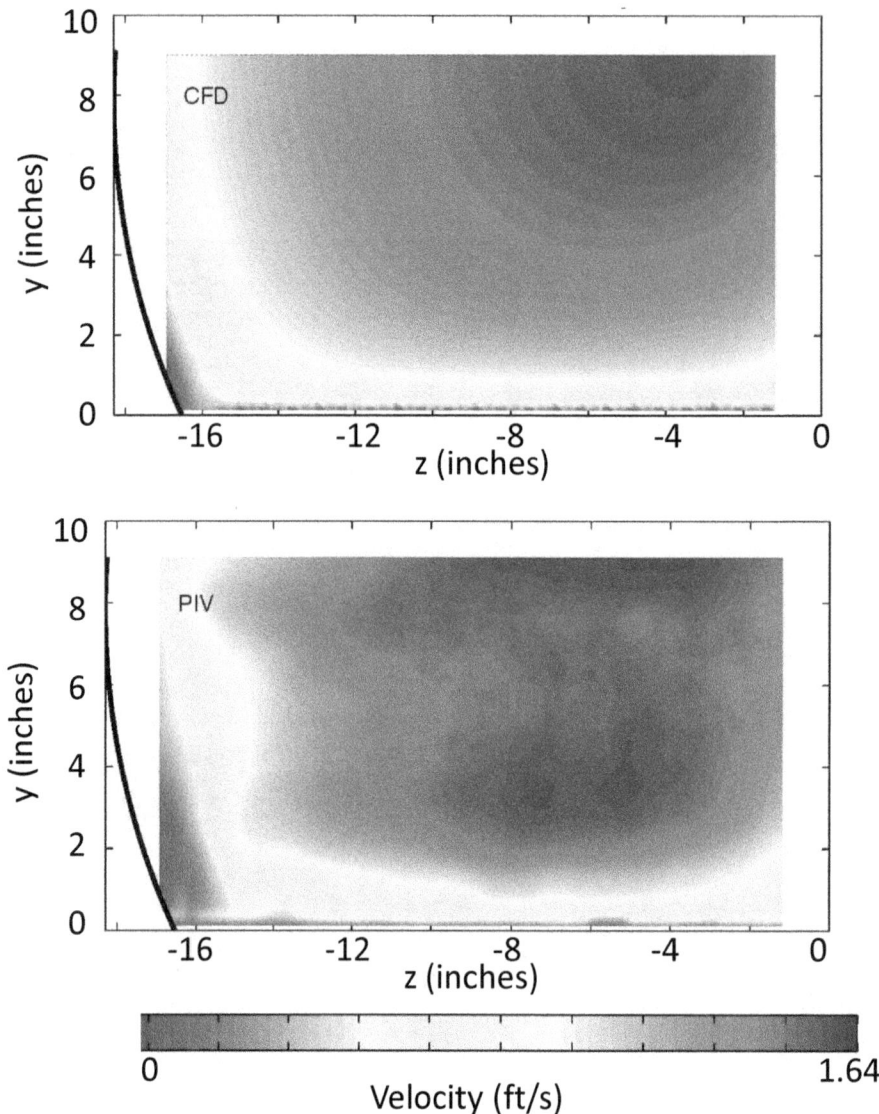

Figure 45. Illustration. Comparison of CFD and PIV results for run F3H00V2D3.

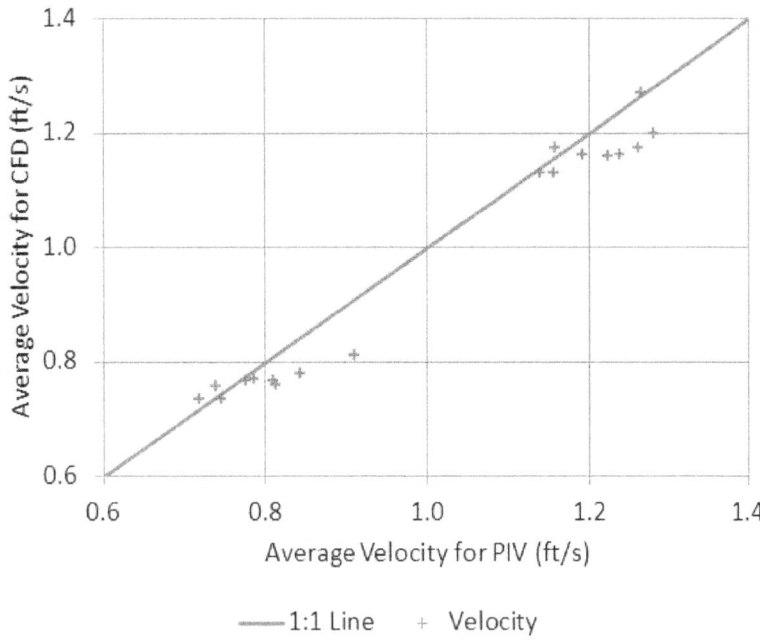

Figure 46. Graph. Comparison of CFD and PIV mean velocities.

Vertical and Horizontal Slice Comparisons

Because the proposed methodology is based on depth-averaged vertical slices, comparisons between the methods by vertical slices were completed. Because the ADV probe cannot measure reliably within a relatively large area near the walls, the comparison is primarily between the PIV measurements and CFD results. However, ADV data in the area where the probe provided good readings are included for comparison.

The error in measurements, for either vertical or horizontal slices, is defined in figure 47:

$$RMSE_x = \sqrt{\frac{\sum (V_{x,i,1} - V_{x,i,2})^2}{n_x}}$$

Figure 47. Equation. Slice RMSE.

Where:
$RMSE_x$ = Root mean square difference in velocity for the xth slice, ft/s.
$V_{x,i,1}$ = Velocity at the ith point in the xth slice as computed by CFD, ft/s.
$V_{x,i,2}$ = Velocity at the ith point in the xth slice as measured by PIV, ft/s.
n_x = Number of velocity measurements in the xth slice.

A relative percentage error is used as a normalized measure and is defined in figure 48:

$$RE_x = \frac{RMSE_x}{V_{x,max} - V_{x,min}}$$

Figure 48. Equation. Relative error in the xth slice.

Where:

RE_x = Relative error in the xth slice, dimensionless.
$V_{x,max}$ = Maximum point velocity measured (PIV) in the xth slice, ft/s.
$V_{x,min}$ = Minimum point velocity measured (PIV) in the xth slice, ft/s.

Figure 49 shows a vertical profile (slice) comparison for flume run F3H00V1D1 (no embedment). The vertical slice shown is taken 3.94 inches from the centerline of the 3-ft culvert. Similarly, figure 50 shows the same comparison for flume run F3H00V2D3.

A width averaged (horizontal slice) comparison was also completed. Figure 51 shows a horizontal profile comparison for flume run F3H00V1D1 (no embedment). The horizontal slice shown is taken 3.94 inches from the bottom of the 3-ft culvert. Similarly, figure 52 shows the same comparison for flume run F3H00V2D3. The relative errors for each of the runs compared are listed in table 9 and range from 7.8 to 18.4 percent.

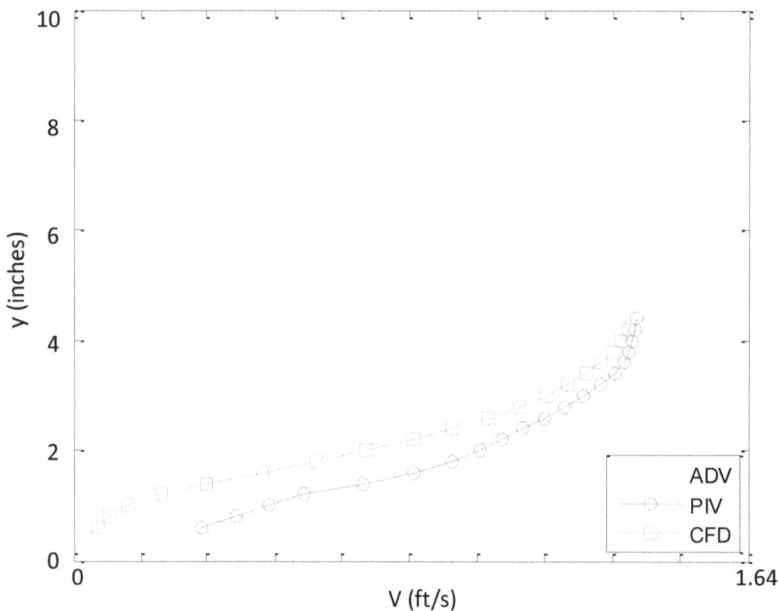

Figure 49. Graph. Vertical slice velocity comparison for flume run F3H00V1D1.

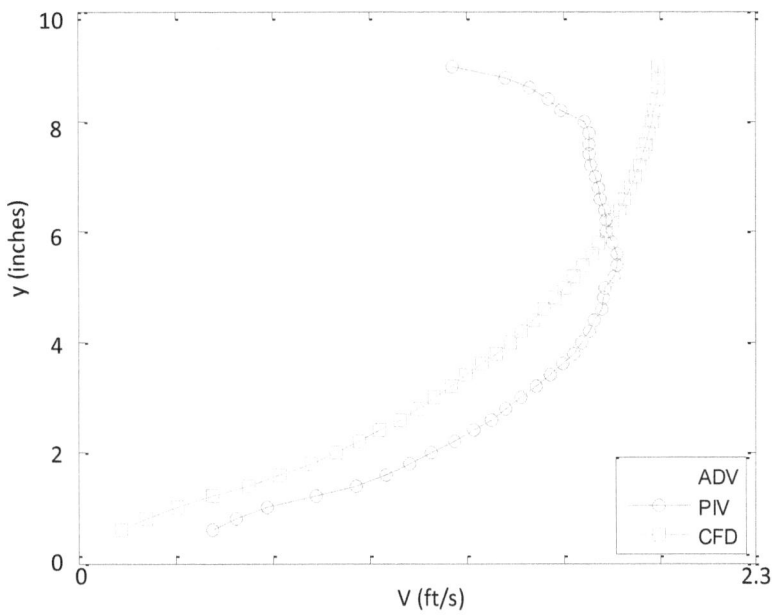

Figure 50. Graph. Vertical slice velocity comparison for flume run F3H00V2D3.

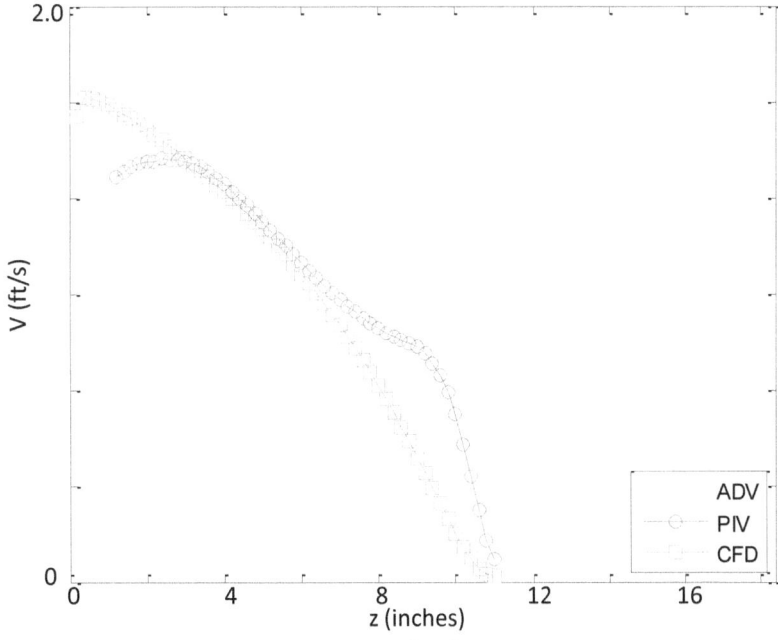

Figure 51. Graph. Horizontal slice velocity comparison for flume run F3H00V1D1.

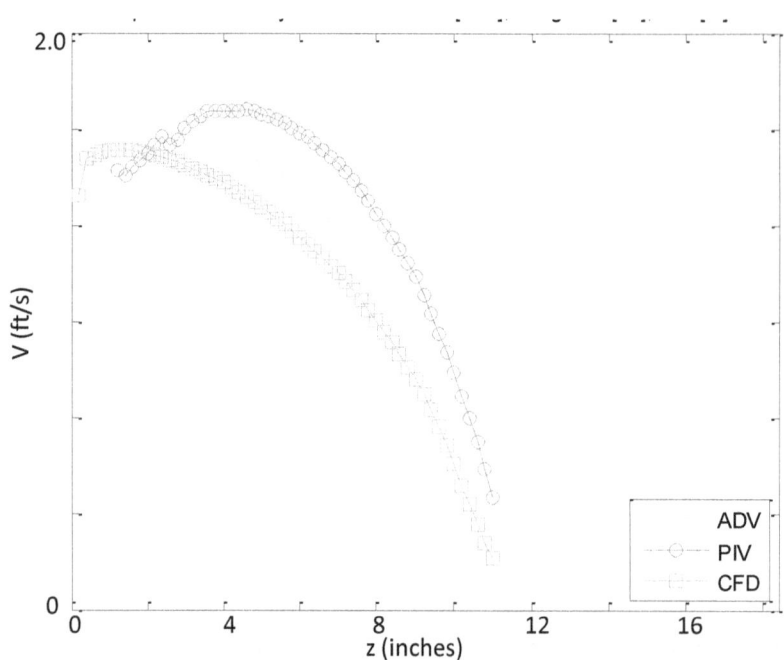

Figure 52. Graph. Horizontal slice velocity comparison for flume run F3H00V2D3.

Table 9. Relative error for slice velocity measurements.

Flume Run ID	CFD Run ID	Embedment (inches)	Velocity (ft/s)	Water Depth (inches)	Vertical Slice Relative Error (percent)	Horizontal Slice Relative Error (percent)
F3H00V1D1	C3H00V1D1	0	0.71	4.5	16.5	11.9
F3H00V1D2	C3H00V1D2	0	0.71	6	16.4	8.4
F3H00V1D3	C3H00V1D3	0	0.71	9	12.2	15.8
F3H00V2D1	C3H00V2D1	0	1.1	4.5	13.8	16.1
F3H00V2D2	C3H00V2D2	0	1.1	6	14.0	7.8
F3H00V2D3	C3H00V2D3	0	1.1	9	14.3	18.4

Sources of Difference

The CFD data are in good agreement with experimental measurements using PIV. The differences may be attributed to one or more of the following reasons:

1. A trumpet-shaped inlet with a honeycomb flow straightener was used in combination with flume tilt and flap gate angle to obtain a nearly uniform flow condition at the test section in the flume where velocity data were obtained. Because fully developed flow requires a very long channel, some error is expected when it is compared with the fully developed flow from the cyclic boundary condition in CFD.

2. Error in the discharge measured by the magnetic flow meters might contribute to a small part of the total error.

3. Explicit assumptions used in the CFD modeling and implicit assumptions embedded in the commercial CFD codes may contribute to differences.

4. Interpolation errors when assigning velocities to the sampling grid may have occurred.

5. Other minor experimental errors.

It is also noted that the CFD measurements are generally taken in the trough section of the corrugation rather than on the inner section of the corrugation. This approach results in a larger cross-sectional area than would be expected using nominal culvert dimensions because the trough extends beyond the nominal pipe diameter. With the larger cross-sectional area and a given flow, the average velocity is somewhat smaller than would be nominally computed.

In the following analyses, the magnitude of the velocity vectors at each point is taken as the velocity for that point. In most of the culvert cross-section, the direction of the vector is aligned with the longitudinal flow. However, in the trough area, the velocity vectors point in many directions because of the turbulence caused by the corrugations. The magnitude of these vectors is small in comparison with those aligned with the flow.

SYMETRICAL HALF VERSUS FULL CULVERT MODELS

Although useful for comparing CFD modeling with physical modeling in the flume, the symmetrical half sections are not appropriate for evaluating the velocity distribution across a culvert section. Figure 53 presents a comparison of the half-section and full-section runs for the 6-inch depth with the 0.71 ft/s velocity for the 0- and 30-percent embedment conditions. At the centerline, where there is a wall in the half section, the velocity is significantly reduced for the half-section runs compared with the full-section runs. Other flow conditions behave in a similar manner. Only full-section runs are evaluated further for development of a design methodology.

SCALING TO LARGER MODELS

With the CFD modeling, there is an opportunity to evaluate many more and larger models than is feasible in the laboratory flume. Test runs with 6- and 8-ft diameter corrugated metal pipes were conducted as previously described in test matrices CFD3 and CFD4, respectively. The sensitivity to parameters such as size and corrugation dimensions is tested.

6-ft Diameter Model

The 6-ft diameter corrugated metal pipe is scaled up from the 3-ft model based on the Froude number and a 2:1 length ratio. The laboratory model culvert corrugation is 3 inches by 1 inch, while the 6-ft model has 6- by 2-inch corrugations. The embedment and water depths were all twice the size of the flume model. Two velocities were tested in the flume, 0.71 ft/s and 1.1 ft/s, which correspond to the larger scale model velocities of 1.0 and 1.56 ft/s, respectively. The laboratory bed material D_{50} equals 0.472 inches; therefore, the larger-scale bed material D_{50} equals 0.944 inches. For the 6-ft diameter model, the base size of the meshing increased to twice that for the 3-ft culvert in the CFD modeling. Because the model geometry is larger than the previous model and simulation speed is influenced by a larger mesh, run times were extended.

Because the proposed fish passage design is based on the depth-averaged velocities in various slices across the width of the culvert, the results of the original 3-ft diameter model and the scaled 6-ft diameter model are compared. To adjust for scale, the velocity is normalized by average velocity, and position is normalized by the culvert diameter. Figure 54 through figure 56 illustrate the comparison for embedment elevations of 0, 15, and 30 percent of the culvert diameter, respectively.

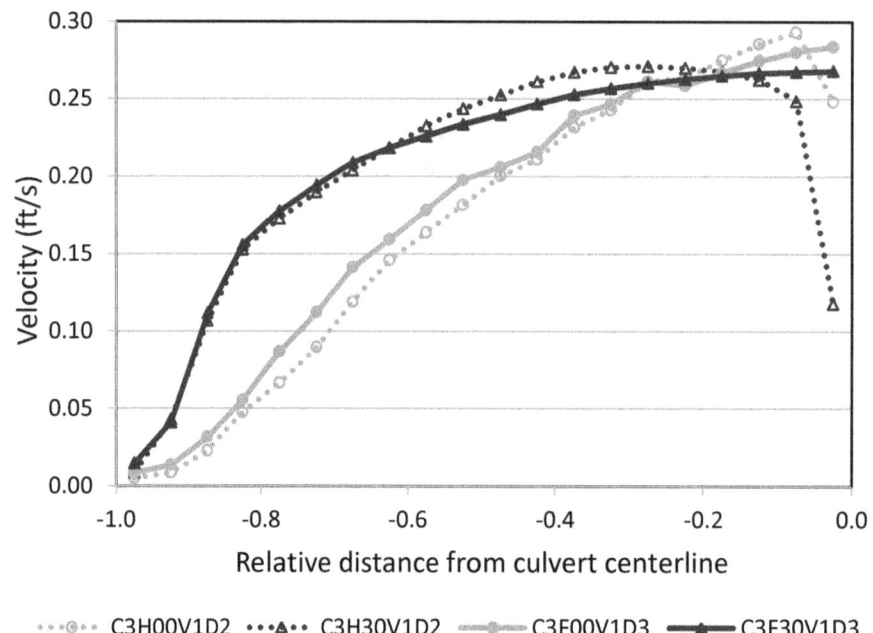

Figure 53. Graph. Selected comparison of half- and full-section CFD runs.

8-ft Diameter Model

The CFD culvert model was also applied to an 8-ft diameter corrugated metal pipe. Many of the length parameters were scaled up from the 3-ft CMP tests. Flow depths were maintained at 8, 16, and 25 percent of the culvert diameter equaling to 7.7, 15.4, and 24 inches, respectively. Embedment was tested at 0, 15, and 30 percent of the culvert diameter.

The velocities were chosen to be within the range of concern for fish passage at this scale, not based on Froude number similitude. Two velocities were tested for each geometric configuration: 1.0 and 3.0 ft/s. The roughness elements—corrugations and bed material—were the same as those used for the 6-ft model. The bed was modeled to simulate gravel with a D_{50} equal to 0.944 inches, and the corrugations were 6 by 2 inches.

The resulting velocity distributions are displayed in figure 57, figure 58, and figure 59. These plots demonstrate that there are steep velocity gradients close to the bed and walls, which is characteristic of turbulent flow with fully developed boundary layers.

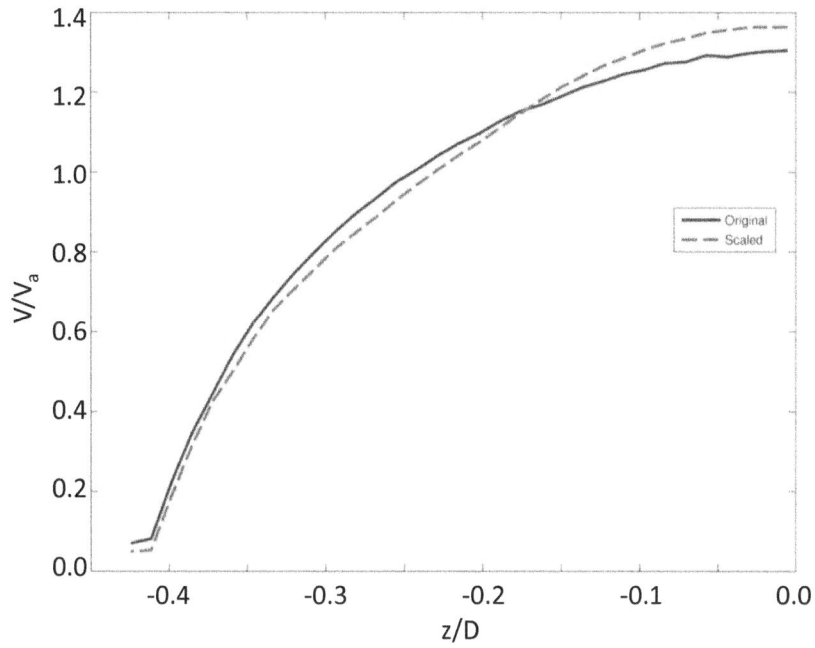

Figure 54. Graph. Comparison of 6- and 3-ft models with no embedment.

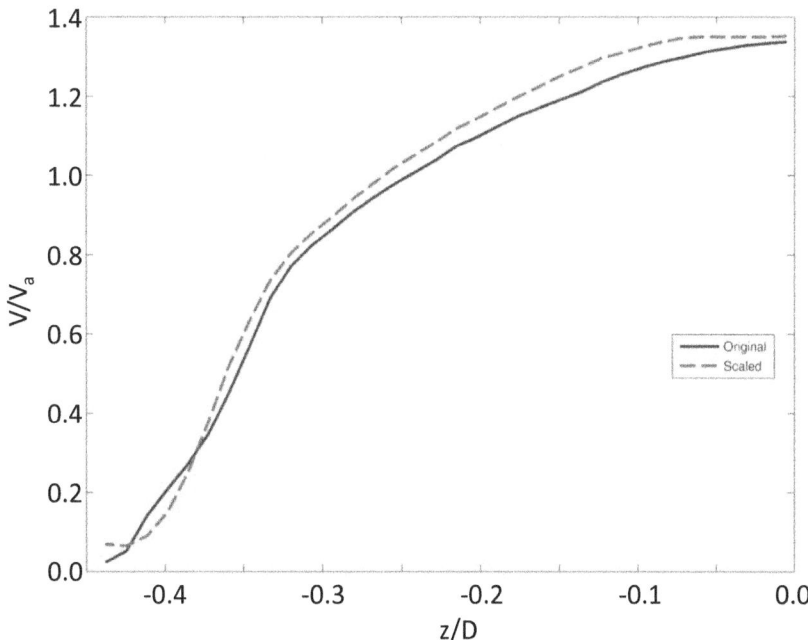

Figure 55. Graph. Comparison of 6- and 3-ft models with 15-percent embedment.

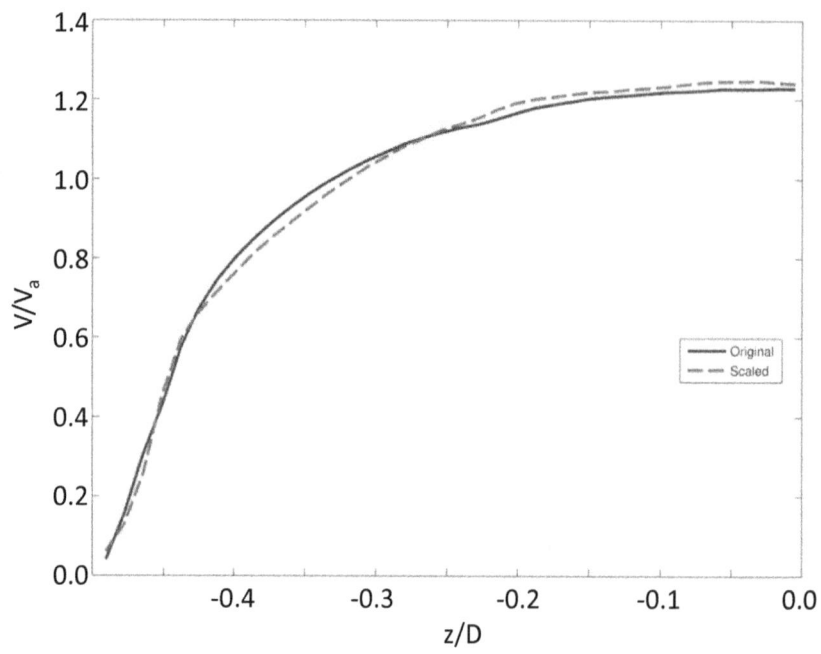

Figure 56. Graph. Comparison of 6- and 3-ft models with 30-percent embedment.

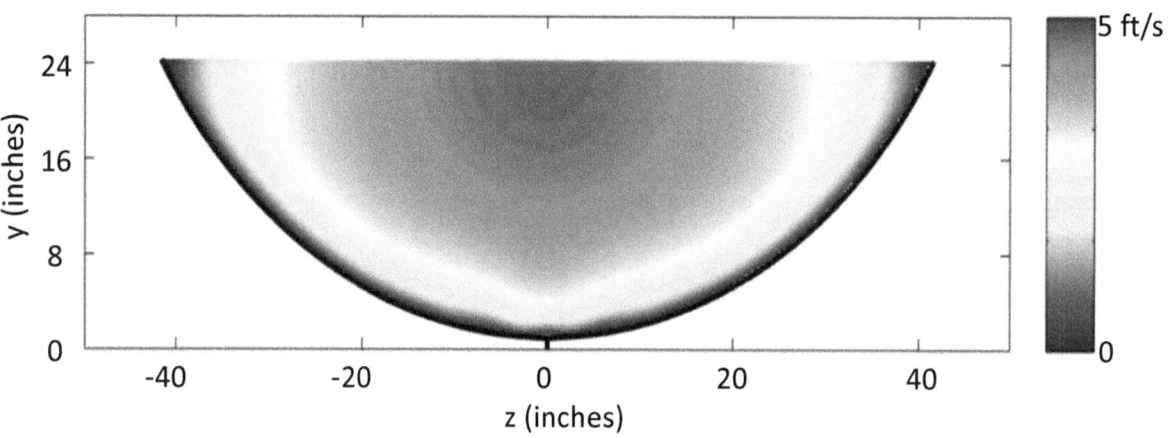

Figure 57. Illustration. Velocities for run C8F00V2D3 (no embedment).

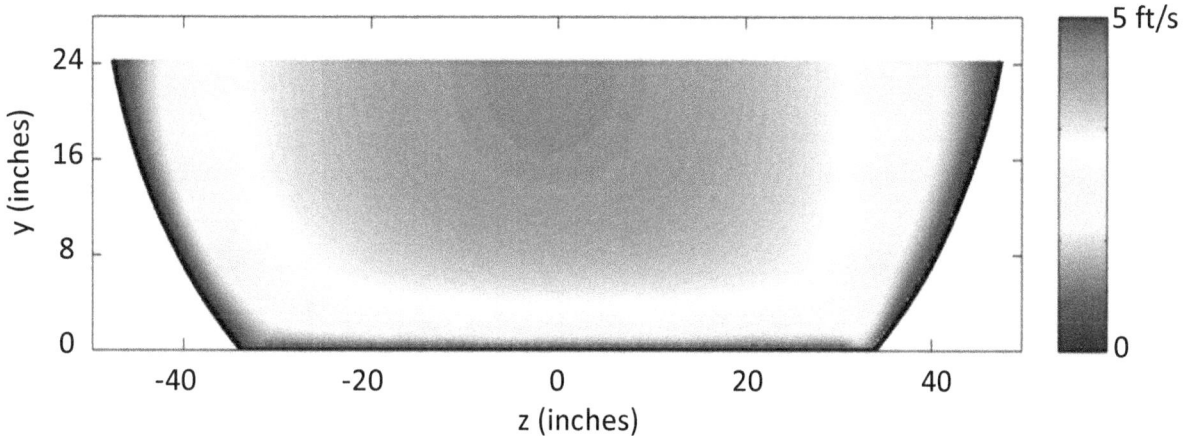

Figure 58. Illustration. Velocities for run C8F15V2D3 (15-percent embedment).

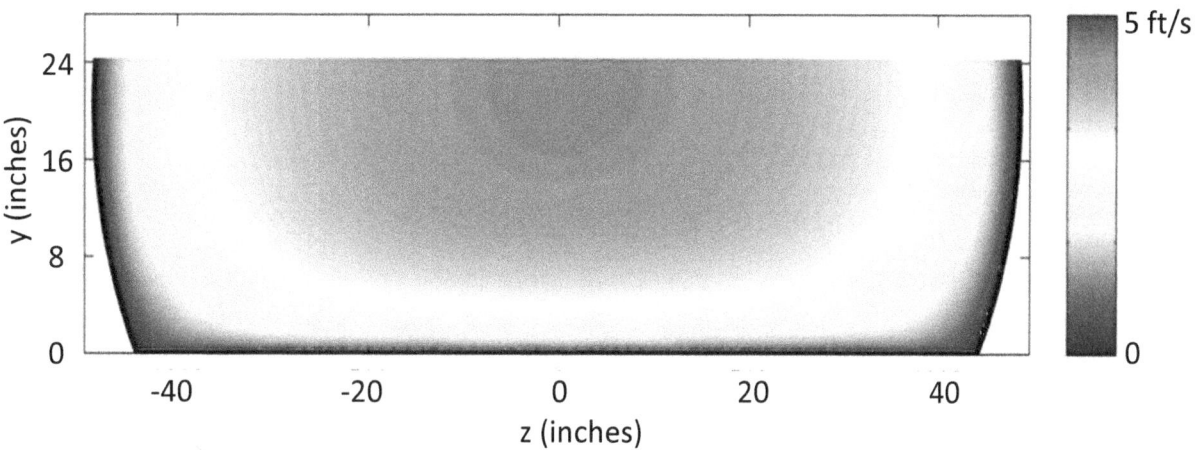

Figure 59. Illustration. Velocities for run C8F30V2D3 (30-percent embedment).

CHAPTER 6. ANALYTICAL DEVELOPMENT

To achieve the research goal of developing design tools to estimate the velocity distribution in an embedded culvert, a means of characterizing the velocity distribution is needed. House et al. used regression techniques to estimate the percentage of a stream cross-section less than or equal to a given velocity.[9] This work is limited to the relatively narrow range of observed conditions used in the regression. Chiu and Chiou and then Chiu describe a mathematical model for velocity distribution that may have broader application and was adapted for this research.[30,31]

VELOCITY DISTRIBUTION MODEL

In the mathematical model described by Chiu and Chiou, the Cartesian coordinate system is replaced with a curvilinear system that aligns with the equal velocity curves (isovels) in a channel as shown in figure 60 and figure 61.[30] In this coordinate system, Y and Z are replaced by ξ and η, where the ξ curves represent the isovels. y_{max} represents the flow depth at the deepest point in the channel, typically the centerline, and B_i represents the top width of the flow to the left (i = 1) and right (i = 2) of the centerline, allowing for consideration of an asymmetrical channel. The difference between the two figures is the value of the parameter ε. In figure 60, it is non-negative (not shown), indicating that velocity continues to increase at the centerline as the water surface is approached. It is negative in figure 61, when the peak velocity is located below the water surface.

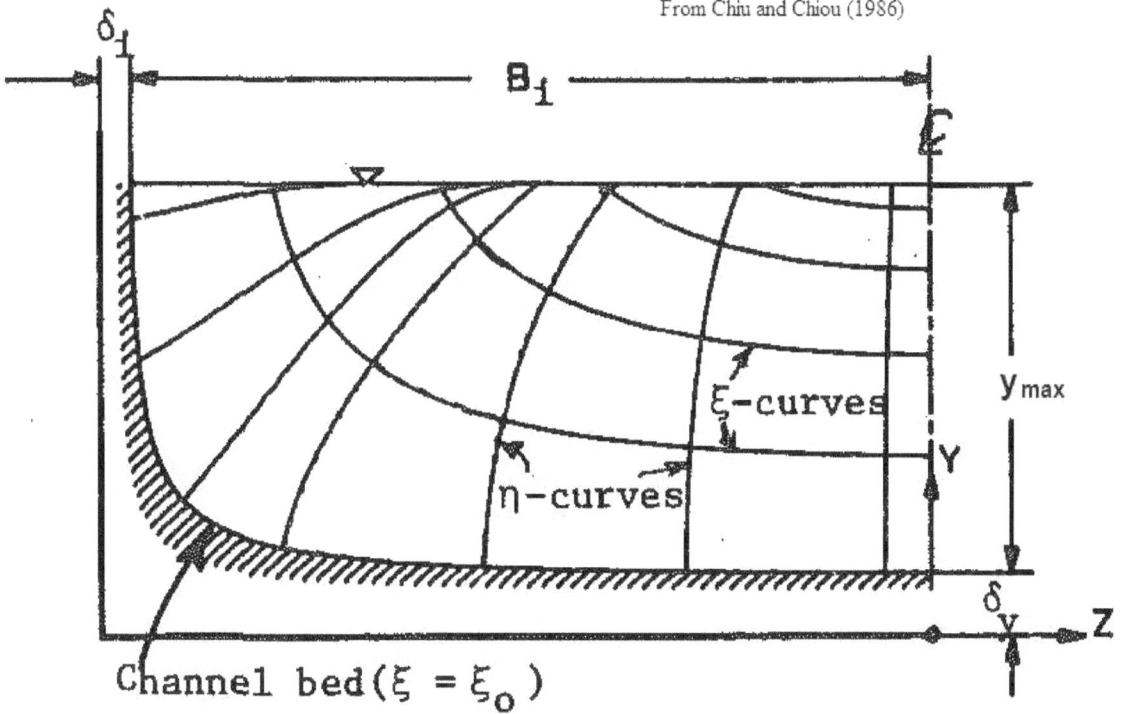

Figure 60. Illustration. Isovel coordinate system ($\varepsilon \geq 0$).

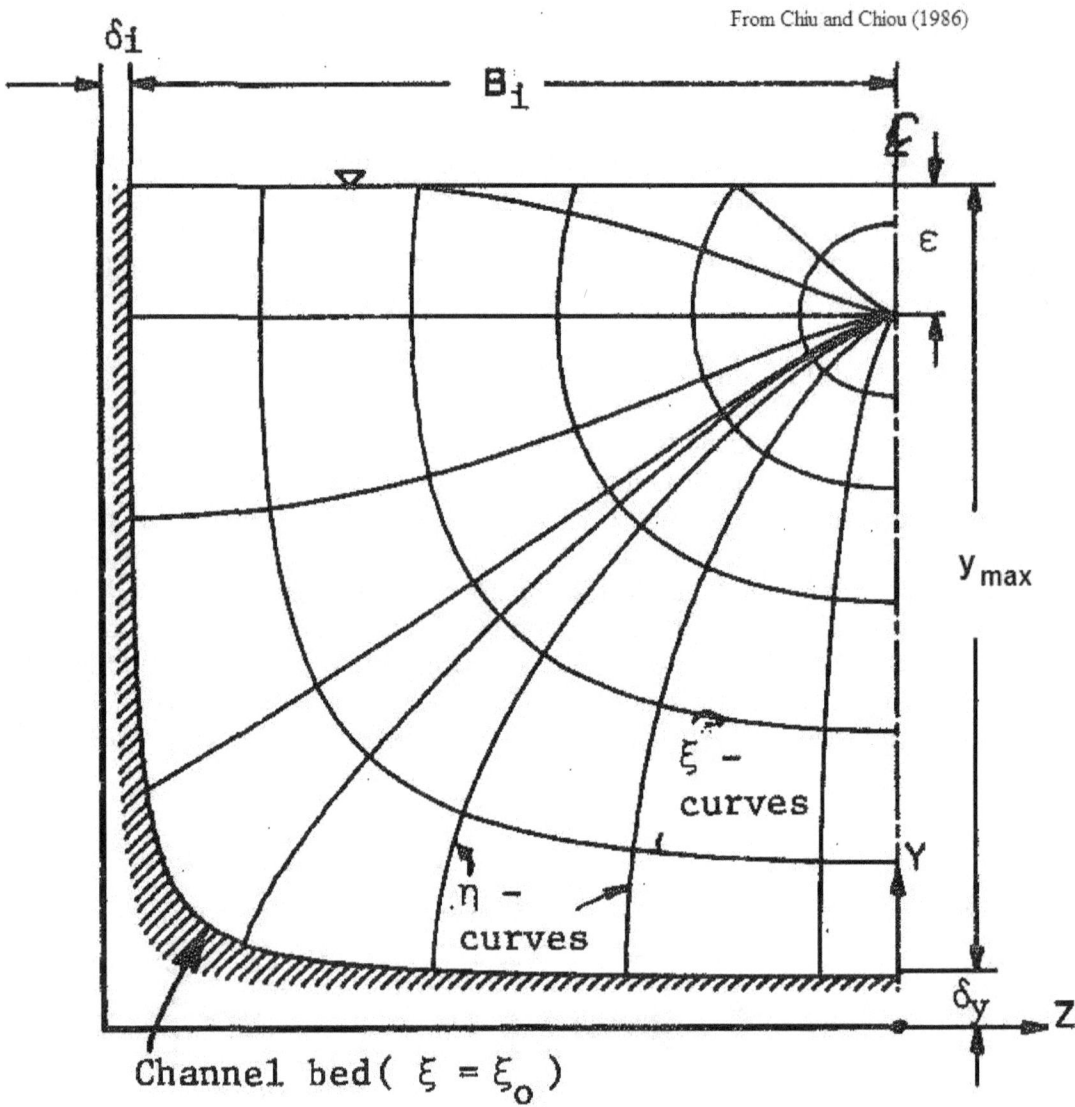

From Chiu and Chiou (1986)

Figure 61. Illustration. Isovel coordinate system ($\varepsilon < 0$).

The primary isovels can be estimated as shown in figure 62:

$$\xi = Y(1-Z)^{\beta_i} \, exp(\beta_i Z - Y + 1)$$

Figure 62. Equation. Isovel equation.

Where:
ξ = Primary isovel coordinate value.
Y = Normalized Cartesian coordinate in the vertical direction.
Z = Normalized Cartesian coordinate in the horizontal direction.
β_i = Velocity distribution parameter.

The normalized Cartesian coordinates are computed from the equations in figure 63 and figure 64.

54

$$Y = \frac{y + \delta_y}{y_{max} + \delta_y + \varepsilon}$$

Figure 63. Equation. Normalized _Y_ coordinate.

Where:

y = Vertical distance, ft.

δ_y = Velocity distribution parameter as shown in figure 60 and figure 61, ft.

y_{max} = Water depth at the y-axis (culvert centerline), ft.

ε = Velocity distribution parameter as shown in figure 60 and figure 61, ft.

$$Z = \frac{|z|}{B_i + \delta_i}$$

Figure 64. Equation. Normalized _Z_ coordinate.

Where:

z = Horizontal distance from the culvert centerline, ft.

δ_i = Velocity distribution parameter as shown in figure 60 and figure 61, ft.

β_i = Transverse distance on the water surface to the left (i = 1) and right (i = 2), ft.

The parameters δ_y, δ_i, and β_i primarily control the shape of the zero-velocity isovel and the isovels near the bed and sides of the channel. The parameter ε primarily influences the higher isovels closer to the surface in the center of the channel.

Figure 65 presents Chiu's relation for computing the velocity at any point in the channel based on the maximum velocity in the channel and the ξ value.[31]

$$\frac{V}{V_{max}} = \frac{1}{M} ln\left[1 + \left(e^M - 1\right)\frac{\xi - \xi_0}{\xi_{max} - \xi_0} \right]$$

Figure 65. Equation. Point velocity.

Where:

V = Point velocity for the ξ isovel, ft/s.

V_{max} = Maximum point velocity in the channel cross-section, ft/s.

M = Velocity distribution parameter.

ξ_o = Minimum value for the ξ coordinate.

ξ_{max} = Maximum value for the ξ coordinate.

Using the equations in figure 62 through figure 65, the velocity at any point in the channel can be estimated if the velocity distribution parameters can be determined. Zhai applied regression techniques to the CFD modeling results for the 3-ft culvert presented earlier to estimate M, ε, β_i, δ_i, and δ_y.[32] However, Zhai's analysis resulted in parameter values inconsistent with the physical meaning of several of the parameters. Zhai also did not provide any guidance on estimating these parameters for design purposes.

Figure 66 through figure 68 illustrate the depth-averaged velocity distributions from the CFD modeling for no embedment, 15-percent embedment, and 30-percent embedment, respectively, for the V1 conditions. These figures are the normalized such that the abscissa is the distance from the culvert centerline, z (negative indicates measurement to the left of the centerline), relative to the water surface width from the centerline to the left edge of the water surface, B_l. The ordinate axis is the vertically averaged velocity at that location, V, relative to the average velocity in the culvert, V_a. Similar patterns were observed for the higher V2 velocity.

Immediately apparent is the greater profile variation for the no-embedment runs. Conversely, the velocity distributions are virtually identical for the 30-percent embedment runs where the channel cross-sections approximate a rectangular shape. Figure 69 summarizes a cross-comparison for the intermediate depth, D4, illustrating the differences in the velocity distributions with embedment. Reasonable estimates of the velocity distribution parameters are necessary to capture these variations.

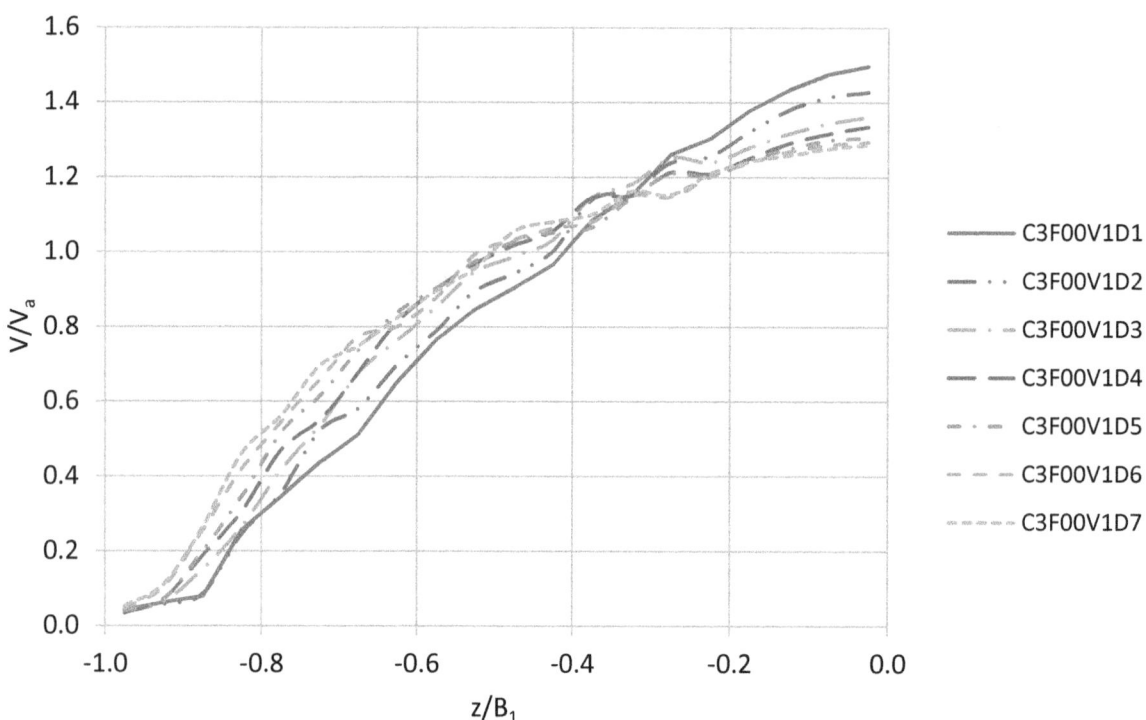

Figure 66. Graph. Vertical slice velocity for no embedment (V1).

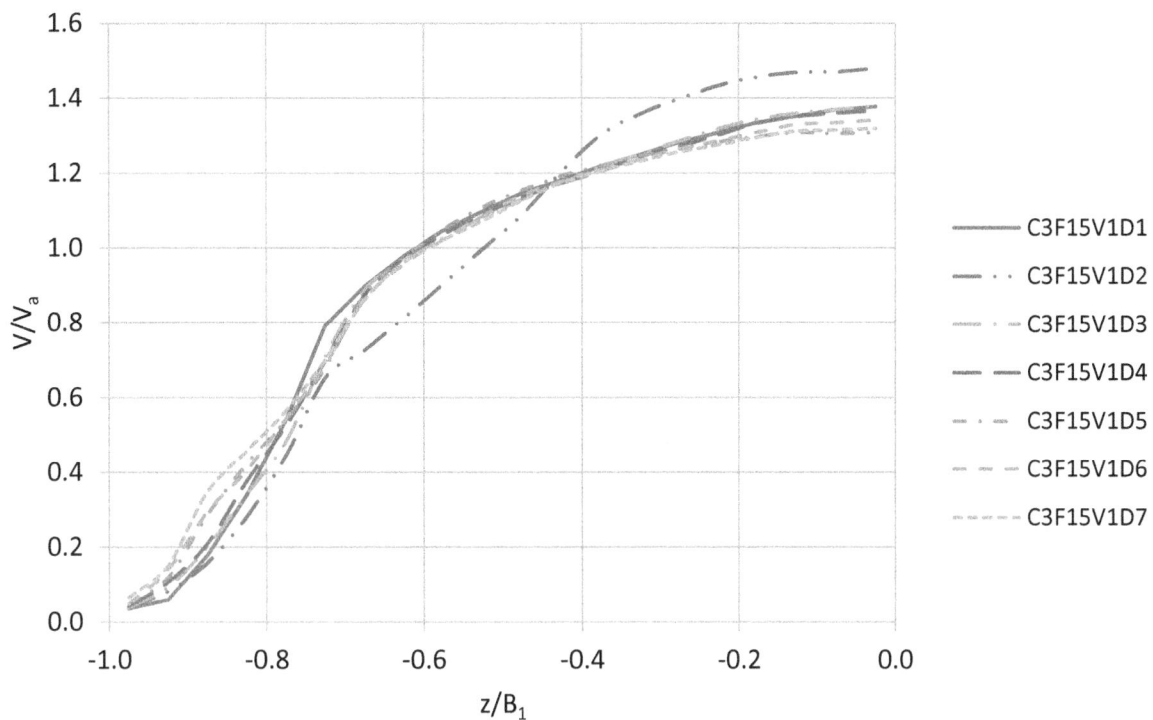

Figure 67. Graph. Vertical slice velocity for 15-percent embedment (V1).

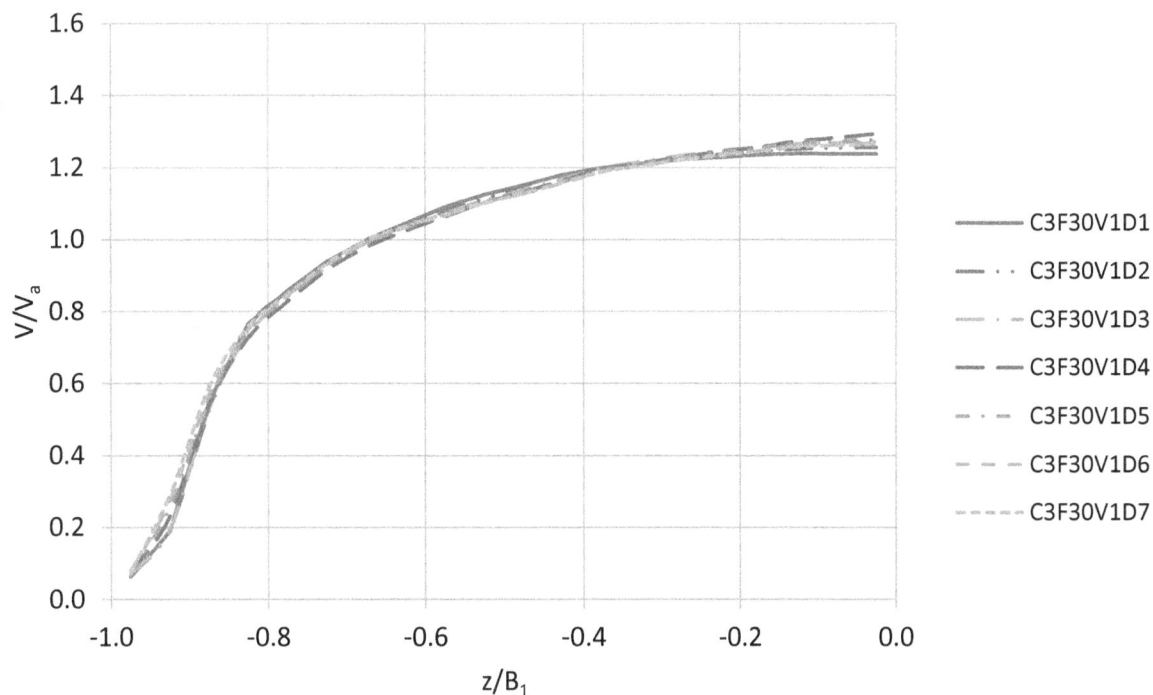

Figure 68. Graph. Vertical slice velocity for 30-percent embedment (V1).

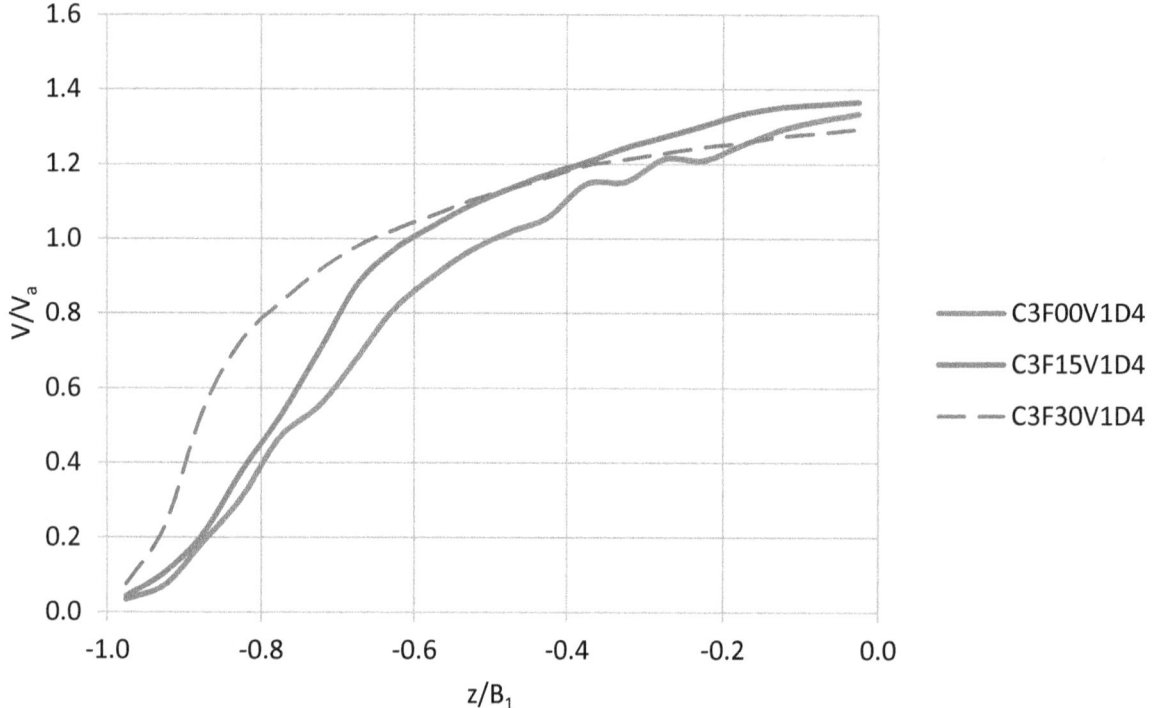

Figure 69. Graph. Vertical slice comparison (V1D4).

DESIGN EQUATION DEVELOPMENT

A structured approach to parameter estimation is conducted while respecting the physical meaning of the parameters. Parameters are estimated based on the 3-ft culvert CFD runs and validated on the 6- and 8-ft culvert CFD runs. Because the parameters M and β_i more fundamentally affect the computation of the isovels, these two parameters are addressed first, starting with M, the velocity distribution parameter related to the maximum point velocity in the cross-section. Once the parameters are fitted to the CFD data, design equations to estimate those parameters are developed.

Maximum Velocity

The maximum point velocity from each CFD run was noted. Zhai provided a relation between the parameter M and the mean and maximum velocities in the channel as shown in figure 70.[32] M may be estimated if the maximum velocity in the channel can be estimated. (Average velocity will be known using standard hydraulics techniques when designing a culvert.)

$$\frac{V_a}{V_{max}} = \frac{e^M}{e^M - 1} - \frac{1}{M}$$

Figure 70. Equation. Relation of M to velocity.

Where:
V_a = Average velocity in the channel cross-section, ft/s.
V_{max} = Maximum velocity in the channel cross-section, ft/s.

M = Velocity distribution parameter.

Several hypotheses for estimating the ratio of average discharge to maximum discharge were tested, including functions of various dimensionless ratios capturing geometric and hydraulic quantities:

- Hydraulic radius divided by average depth, R_h/y_a.

- Top width squared divided by cross-sectional area, T^2/A.

- Maximum top width (half section) divided by the average flow width (half section), B_1/B_{avg}.

- Shear velocity divided by average velocity, u_*/V_a.

- Unit discharge, Q/T.

- Froude number, Fr.

- Hydraulic radius to the one-sixth power divided by n, $R_h^{1/6}/n$.

(Top width, T, is equal to $B_1 + B_2$. For symmetrical cross-sections, $B_1 = B_2$ and T equals $2B_1$.) (Average flow width, B_{avg}, equals $A/(2y_{max})$.)

The ratio of maximum velocity to average velocity did not exhibit a single relationship to any of the variables noted above for all embedment levels. However, one of the strongest patterns was indicated when plotted against the ratio of T^2/A. Figure 71 displays this relation, distinguishing the run series by velocity and embedment. However, the relation is quite different for the C3F00V1 and C3F00V2 series (no embedment) compared with the embedded series with 15-percent (C3F15V1 and C3F15V2) and 30-percent (C3F30V1 and C3F30V2) embedment. A consequence of the embedment is it shifts the flow cross-section wetted perimeter from a circular arc to an almost rectangular shape.

Inspection of figure 71 suggests that each embedment level separately follows an equation of the form: $Y = aX^b + c$. After exploring several alternatives, the ratio B_1/B_{avg} was selected to determine the equation form for estimating the maximum velocity ratio. Figure 72 shows that variation in the maximum velocity ratio is strongly influenced by B_1/B_{avg} and embedment.

These analyses led to the relation for estimating the maximum velocity ratio based only on channel geometry as shown in figure 73. With the equation in figure 73, the parameter M can now be generated using the equation presented in figure 70.

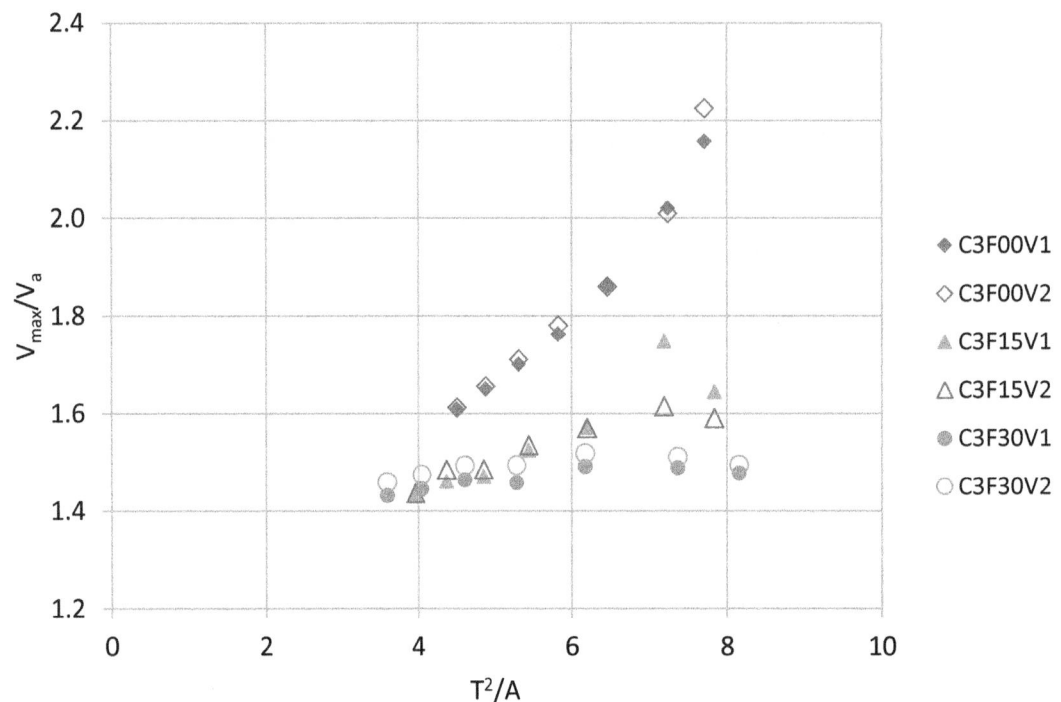

Figure 71. Graph. Variation of V_{max} ratio with T^2/A.

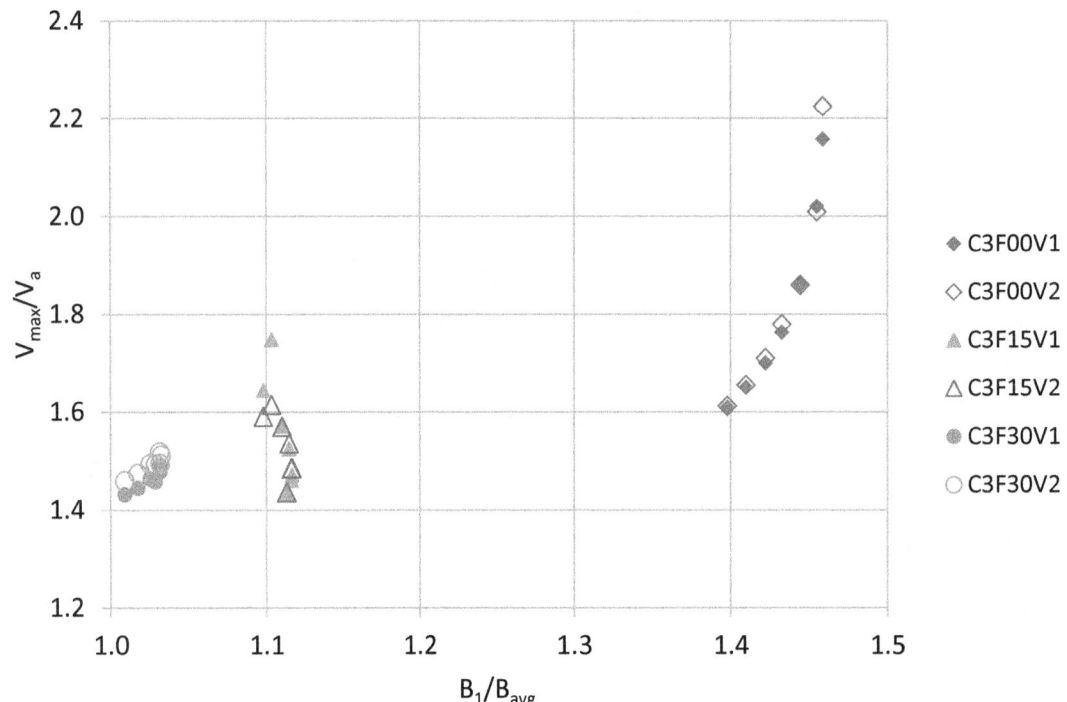

Figure 72. Graph. Variation of V_{max} ratio with B_1/B_{avg}.

$$\frac{V_{max}}{V_a} = \left[0.047 - 0.029\left(\frac{B_1}{B_{avg}}\right)\right]\left[\frac{T^2}{A}\right]^{\left[-2.7+3.6\left(\frac{B_1}{B_{avg}}\right)\right]} + 1.38$$

Figure 73. Equation. Estimation of V_{max}/V_a.

Where:
V_a = Average velocity in the channel cross-section, ft/s.
V_{max} = Maximum point velocity in the channel cross-section, ft/s.
B_{avg} = Average width for the symmetrical culvert half section, ft.
B_1 = Maximum top width for the symmetrical culvert half section, ft.
T = Top width for the full section, ft.
A = Cross-section area for the full section, ft^2.

Figure 73 assumes that the parameters T and B_1 increase as the water surface elevation increases. However, the test matrix includes runs where the two deepest depths with the 30-percent embedment have top widths less than the maximum culvert width (1.1 percent or less). Therefore, relations based on T and B_1 may also be applied when the water surface elevation exceeds the elevation of the maximum culvert width as long as T is within two percent of the maximum width.

A comparison of the observed CFD values and those computed using the equation is shown in figure 74. The figure indicates good agreement with one apparent outlier from the C3F15V1 series (C3F15V1D2). The RMSE for the 42 runs included in this analysis was 0.046, which is equivalent to 2.9 percent of the average V_{max}/V_a ratio.

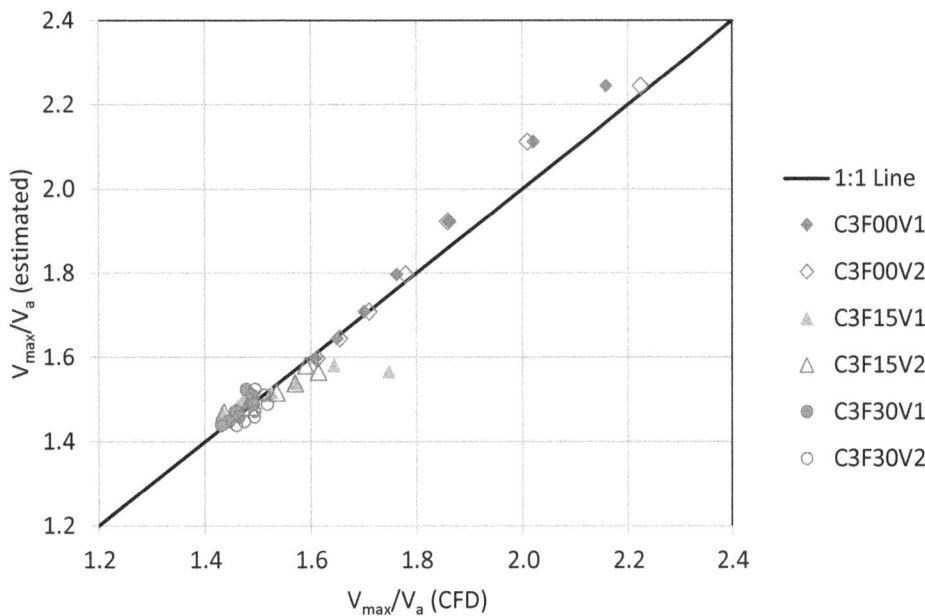

Figure 74. Graph. Calculated versus observed V_{max}/V_a ratios.

Low Velocity Isovel and Maximum Velocity Location Parameters

Estimates of the low velocity isovel parameters (β_i, δ_i, and δ_y,) and the maximum velocity location parameter (ε) were made for each 3-ft CFD run by minimizing the differences between the velocities produced by the CFD model and estimates using the velocity model in figure 65. Because the CFD runs did not provide velocity output at uniform gridded locations for each run and produced many more individual points than could be effectively analyzed, the CFD data were transformed into a grid system. The flow cross-section for each run was divided into 20 equal sections horizontally and 20 equal sections vertically, creating a grid of identically sized rectangles. CFD data were converted to this 20 by 20 grid to form a uniform gridded representation of the velocity profile. Each run had up to 400 velocity points representing the velocity distribution. The actual number was less than 400 because the flow cross-sections are not rectangular. (Because these grids are stair-stepped and completely enclose the circular culvert, estimated cross-sectional areas are overestimated slightly.) In addition to the grid, 20 vertical flow slices were computed to represent the depth-averaged velocity variation across the section.

Because of its greater effect on the results, the low velocity parameter, β_i, was addressed first. While holding δ_i, δ_y, and ε equal to zero and using the calculated value of M, β_i was estimated for each run by minimizing the RMSE of the grid and slice data. (The most robust estimates resulted when the objective function of the fitting process was to minimize the sum of the RMSE of the grid and slice data.) The parameters δ_i, δ_y, and ε were held to zero because odd "optimal" solutions were found for isolated runs when they were allowed to vary simultaneously in the analysis. That is, best-fits that did not correspond to physical reality were observed. For most runs, this process had little effect on the outcome.

Fitted estimates of β_i are shown in table 10 (along with the calculated values of M). For the runs with no embedment (C3F00VxDx), β_i declines with flow depth and is slightly higher for the higher-velocity runs (V2). The pattern differs for the runs with embedment at 15 percent of the culvert diameter (C3F15VxDx). For these runs, β_i increases with flow depth and is slightly higher for the higher-velocity runs. For the runs with embedment at 30 percent (C3F30VxDx), β_i is constant at about 1.3. Again, it appears that as the wetted perimeter geometry changes from a circular arc to a rectangular shape, the velocity distribution parameters change accordingly.

Several of the independent variables previously identified were evaluated as possible predictors of β_i. Figure 75 displays the relation with the ratio B_l/B_{avg}. There does appear to be another influence on β_i in addition to this ratio, especially for the no-embedment and 15-percent embedment runs. Other variables that explained other influence were not identified, perhaps because there is insufficient variation in the range of variables within the 42 CFD runs. This may be further investigated but was not within the scope of this research project. The recommended equation for estimating β_i is shown in figure 76.

Figure 77 summarizes the comparison of the fitted values for β_i from the CFD and the estimated values using the equation in figure 76. The RMSE for the 42 runs included in this analysis was 0.12, which is equivalent to 6.3 percent of the average value of β_i.

Table 10. Fitted estimates of velocity distribution parameters for the 3-ft culvert.

Run ID	M	β_i	δ_i (inches)	δ_y (inches)	ε (inches)
C3F00V1D1	-0.45	2.78	0.54	0.00	1.17
C3F00V1D2	-0.06	2.69	0.54	0.00	1.62
C3F00V1D3	0.45	2.63	0.59	0.00	2.59
C3F00V1D4	0.82	2.67	0.67	0.00	3.70
C3F00V1D5	1.08	2.52	0.58	0.00	3.95
C3F00V1D6	1.31	2.48	0.68	0.00	5.37
C3F00V1D7	1.51	2.50	0.72	0.00	6.83
C3F00V2D1	-0.61	2.99	0.33	0.00	0.50
C3F00V2D2	-0.03	2.72	0.45	0.00	1.44
C3F00V2D3	0.45	2.70	0.47	0.00	2.07
C3F00V2D4	0.75	2.73	0.57	0.00	3.02
C3F00V2D5	1.03	2.59	0.53	0.00	3.31
C3F00V2D6	1.28	2.55	0.63	0.00	4.60
C3F00V2D7	1.49	2.57	0.68	0.00	5.89
C3F15V1D1	1.34	1.73	0.00	0.22	0.00
C3F15V1D2	0.88	1.98	0.00	0.46	0.27
C3F15V1D3	1.71	1.97	0.00	0.32	0.00
C3F15V1D4	1.98	2.02	0.00	0.38	0.00
C3F15V1D5	2.35	2.00	0.00	0.52	0.00
C3F15V1D6	2.42	2.12	0.00	0.51	0.00
C3F15V1D7	2.65	2.10	0.00	0.56	0.00
C3F15V2D1	1.61	1.80	0.00	0.13	0.00
C3F15V2D2	1.48	1.89	0.00	0.19	0.05
C3F15V2D3	1.72	1.95	0.00	0.24	0.00
C3F15V2D4	1.92	1.99	0.00	0.28	0.00
C3F15V2D5	2.25	2.09	0.00	0.45	0.00
C3F15V2D6	2.25	2.03	0.00	0.33	0.00
C3F15V2D7	2.61	2.12	0.00	0.44	0.00
C3F30V1D1	2.30	1.29	0.00	0.01	0.00
C3F30V1D2	2.22	1.31	0.00	0.04	0.00
C3F30V1D3	2.21	1.32	0.00	0.09	0.02
C3F30V1D4	2.44	1.40	0.00	0.07	0.05
C3F30V1D5	2.40	1.32	0.00	0.16	0.24
C3F30V1D6	2.54	1.31	0.00	0.19	0.00
C3F30V1D7	2.64	1.31	0.00	0.21	0.00
C3F30V2D1	2.19	1.33	0.00	0.02	0.01
C3F30V2D2	2.08	1.33	0.00	0.07	0.10
C3F30V2D3	2.03	1.34	0.00	0.14	0.19
C3F30V2D4	2.19	1.37	0.00	0.14	0.04
C3F30V2D5	2.19	1.35	0.00	0.25	0.46
C3F30V2D6	2.32	1.34	0.00	0.25	0.08
C3F30V2D7	2.43	1.33	0.00	0.26	0.00

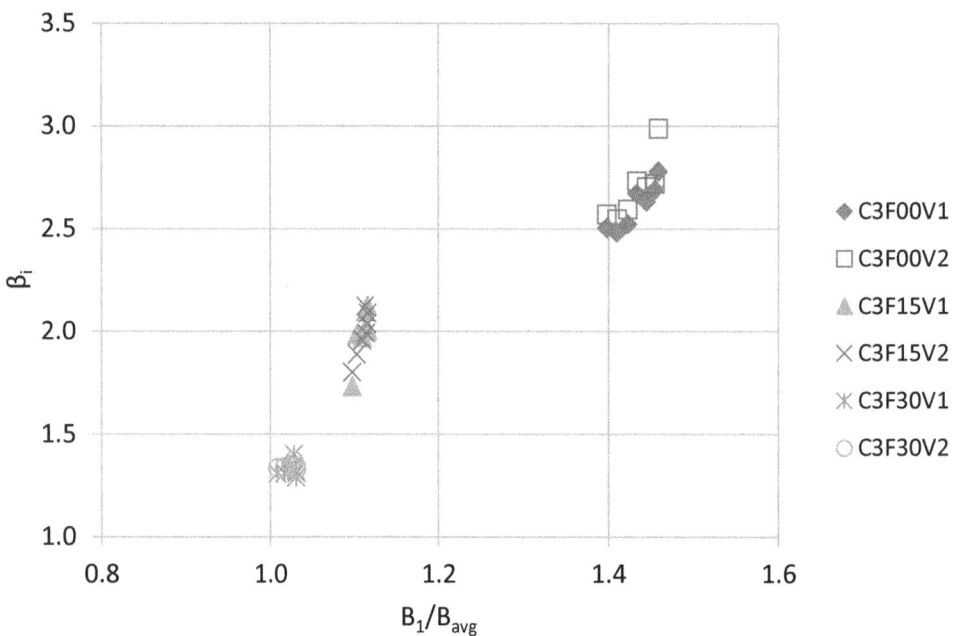

Figure 75. Graph. Variation of β_i with B_1/B_{avg}.

$$\beta_i = 2.56\left(\frac{B_1}{B_{avg}} - 1\right)^{0.49} + 1$$

Figure 76. Equation. Estimation of β_i.

The remaining three velocity distribution parameters, ε, δ_i, and δ_y, were fitted simultaneously, holding the previously determined values of M and β_i fixed. δ_i and δ_y were restricted to non-negative values to be consistent with their physical meaning as shown in figure 60 and figure 61. (When allowed to be negative, the optimization rarely selected negative values for these two parameters.)

Initially, ε was allowed to take on any value because it can be negative, zero, or positive. Because ε represents the location of the maximum velocity in the cross-section (positive values are the distance above the water surface, and negative values are the distance below the water surface), the appropriate value of ε may be validated by inspection of the CFD data. In all CFD runs, the maximum velocity was at the surface or increasing when the surface was reached based on the gridded velocities. Therefore, ε was also restricted to non-negative values.

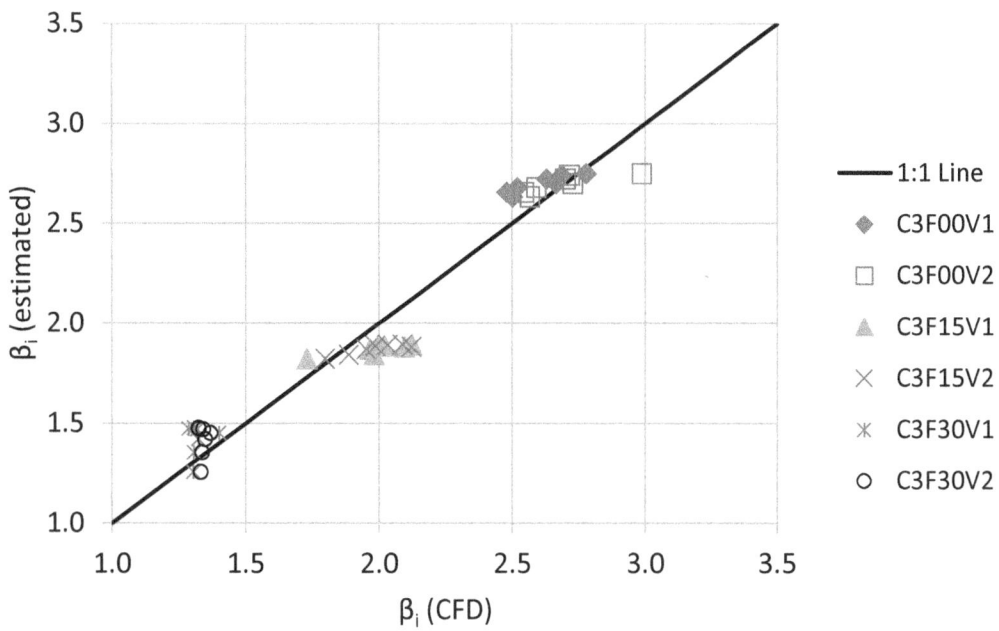

Figure 77. Graph. Estimated versus CFD β_i.

Figure 78 shows a centerline plot of the middle depth (D4) at each embedment level and velocity level from the 3-ft CFD runs. The profile for no embedment differs significantly from the other embedment levels. This differentiation in profile caused by the cross-section changes results in significant differences in the fitted values of the parameters ε, δ_i, and δ_y. The zero embedment cases are the furthest removed from a rectangular-shaped flow cross-section.

Any approach for estimating ε must recognize the differing velocity profiles shown in figure 78. For the no-embedment cases (C3F00), the velocity appears to be increasing slightly as the water surface is approached, suggesting that for these cases, ε should be greater than zero in accordance with the velocity distribution model shown in figure 60. The values reported in table 10 confirm this observation.

Similarly, the velocity profiles for the 15- and 30-percent levels of embedment indicate that maximum velocity is reached at or near the surface, meaning that ε should be zero. For most runs, zero was obtained in the optimization for the C3F15 and C3F30 series. Several runs resulted in small negative values of ε that made only nominal improvements in the fit of the profiles to the CFD results. These values were forced to zero.

The fitted estimates for δ_i and δ_y are also summarized in table 10. For the runs with no embedment, the best fit values of δ_y were all zero, with nonzero values of ε and δ_i. For these runs, using these nonzero values significantly reduced prediction errors for the gridded and slice velocity estimates. Conversely, for the runs with 15-percent and 30-percent embedment, the best fit values of δ_y were all nonzero, with zero values of δ_i in all cases. For these runs, using the nonzero values did not significantly reduce prediction errors for the gridded and slice velocity estimates.

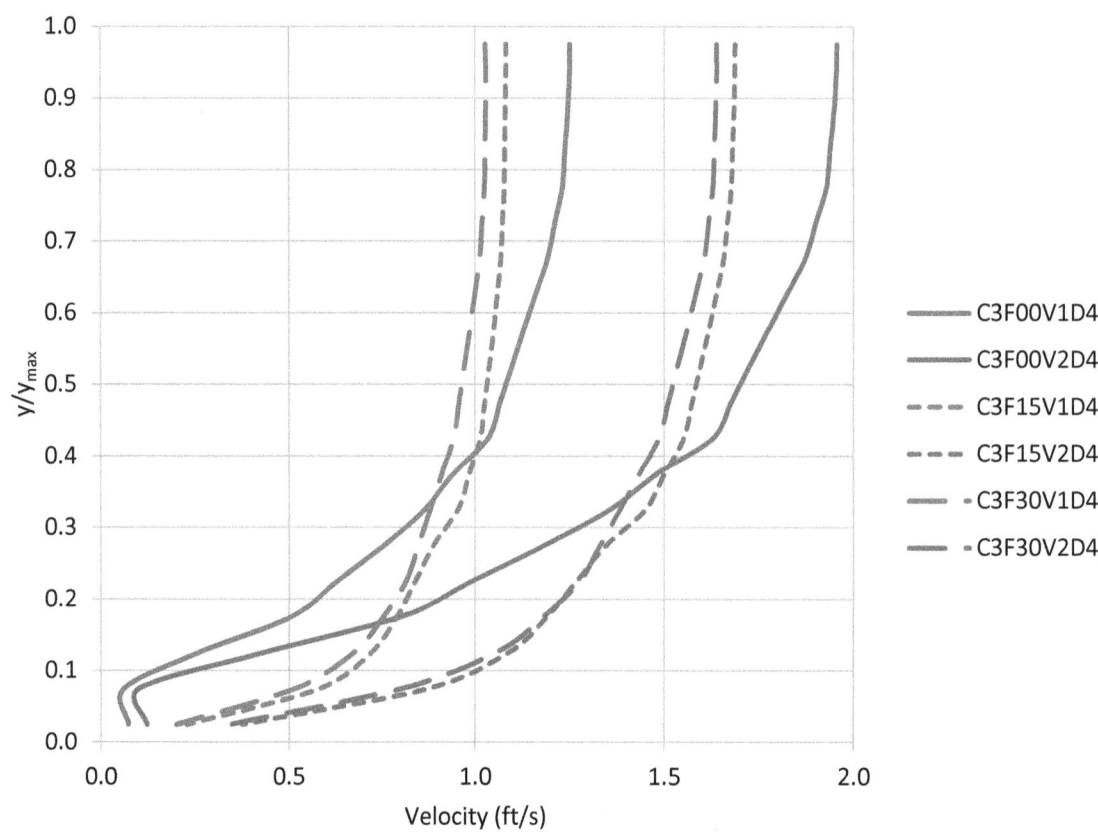

Figure 78. Graph. Velocity profiles for the D4 depth.

For the no-embedment runs, several relations were explored as a basis for estimating ε. Two variables identified for further analysis were the T^2/A ratio (see figure 79) and the Froude number (see figure 80). Both show decreases in ε, normalized by the maximum flow depth, with increases in the respective variable. However, for a given value of the T^2/A ratio, the lower-velocity runs (V1) result in a higher normalized ε, while for a given value of the Froude number, the lower-velocity runs result in a lower normalized ε.

Combining these observations, the behavior of the no-embedment cases may be estimated using the equation in figure 81. For the 15- to 30-percent embedment cases, ε is zero. For more general application, the ratio B_1/B_{avg} is used as the means of distinguishing embedment levels. When B_1/B_{avg} is greater than or equal to 1.2, the equation in figure 81 applies; otherwise, ε is zero.

Figure 82 summarizes the comparison of the fitted values for ε/y_{max} from the CFD and the estimated values using the equation in figure 81. The RMSE for the 14 runs included in this analysis was 0.057, which is equivalent to 13.4 percent of the average value of ε/y_{max}.

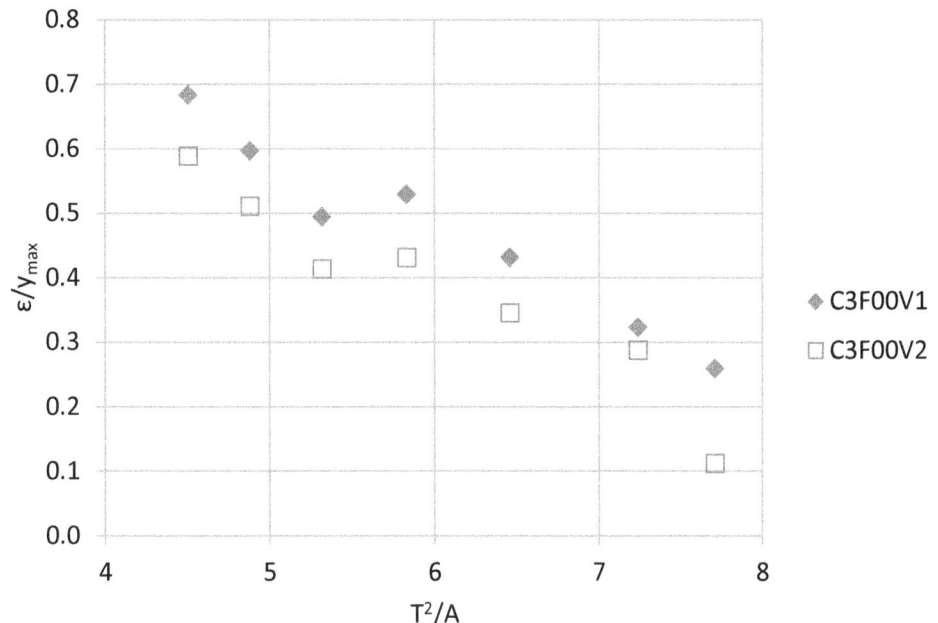

Figure 79. Graph. Variation of ε with T^2/A.

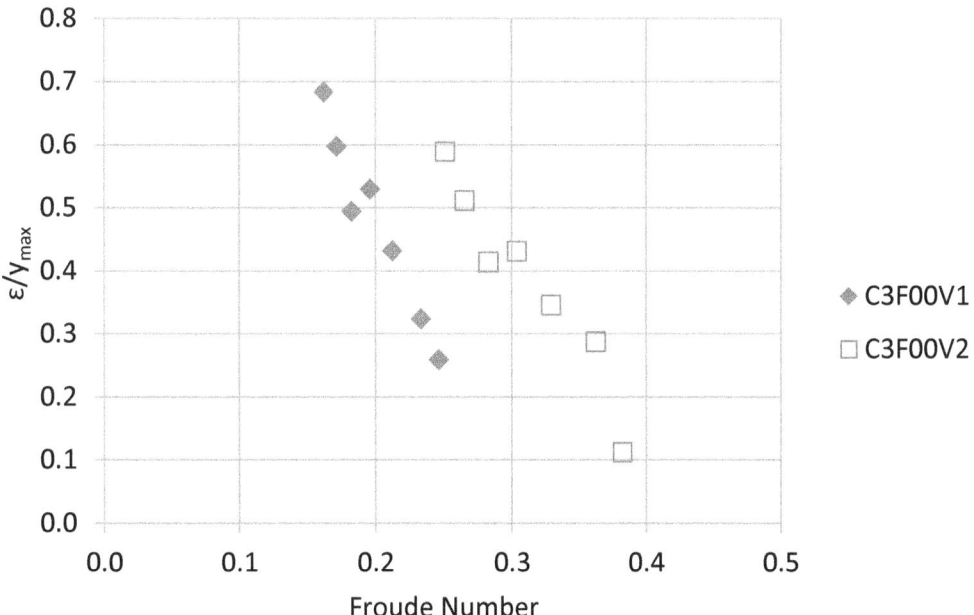

Figure 80. Graph. Variation of ε with Froude number.

$$\frac{\varepsilon}{y_{max}} = 2.56\left(\frac{T^2}{A}\right)^{-1.54}(Fr)^{-0.62}$$

Figure 81. Equation. Estimation of ε for $B_1/B_{avg} \geq 1.2$.

Where:
ε = Velocity profile parameter, ft.

67

y_{max} = Flow depth at the y axis (culvert centerline), ft.
Fr = Froude number based on average cross-section velocity and depth.

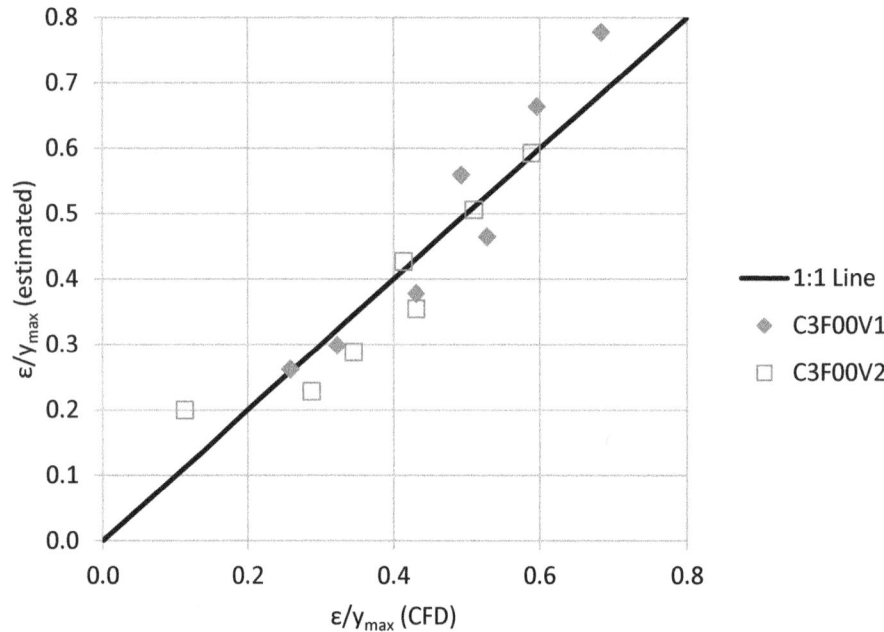

Figure 82. Graph. Estimated versus CFD for ε/y_{max}.

The final two parameters defining the cross-section velocity distribution are the horizontal and vertical offsets. The parameter δ_i represents a horizontal offset adjustment in the computation of Z as defined in figure 64. This parameter is needed to better fit the isovels on the culvert sides in the no-embedment case. The parameter δ_i equals zero in every 15- and 30-percent embedment case. Therefore, when B_1/B_{avg} is less than 1.2, δ_i is set to zero. When B_1/B_{avg} is greater than or equal to 1.2, δ_i is estimated by the equation in figure 83.

$$\delta_i = 0.04B_1$$

Figure 83. Equation. Estimation of δ_i for $B_1/B_{avg} \geq 1.2$.

The parameter δ_y represents a vertical offset adjustment to the computation of Y as defined in figure 63, which is needed to better fit the isovels on the culvert bottom in the embedment cases. δ_y equals zero in every no-embedment case. When B_1/B_{avg} is greater than or equal to 1.2, δ_y is set to zero. For the 15- and 30-percent embedment cases, adjustment of this parameter does not significantly improve the velocity profile fits. However, the equation in figure 84 may be used.

$$\delta_y = a_y y_{max}$$

Figure 84. Equation. Estimation of δ_y for $B_1/B_{avg} < 1.2$.

Where:
δ_y = Velocity profile parameter, ft.
y_{max} = Flow depth at the y axis (culvert centerline), ft.
a_y = Constant dependent on embedment.

The constant, a_y, depends on the degree of embedment as measured by the B_1/B_{avg} ratio and is summarized in table 11. For the zero embedment case, a_y is zero and δ_y is zero. For those cases with a nonzero δ_y, the value of a_y goes down with the ratio of B_1/B_{avg}.

Table 11. Summary of embedment constant a_y.

B_1/B_{avg}	Embedment	a_y
<1.06	30 percent	0.018
< 1.2 and > 1.06	15 percent	0.050
> 1.2	0 percent	0.000

It follows from the design equations for estimating δ_i and δ_y that ξ_0 must equal zero in all design situations addressed in this research report. For the case where B_1/B_{avg} is less than 1.2, δ_y equals zero, and Y (computed from the equation in figure 63) will be zero at the centerline of such a channel at the bed elevation ($z = 0$ and $y = 0$). Similarly, for the case where B_1/B_{avg} is greater than or equal to 1.2, δ_i equals zero, and Z (computed from the equation in figure 64) will be zero at the channel edge at the water surface elevation ($|z| = B_1$ and $y = y_{max}$). Because either Y or Z must equal zero in all cases, there is a value of ξ that will equal zero, according to the equation in figure 62. Therefore, the minimum value of ξ (that is, ξ_0) is zero.

Fitted and Estimated Errors

The RMSE expressed as a percentage of the average velocity for each run were calculated for the velocity distribution grid. Each velocity in the 20 by 20 grid as modeled by CFD was compared to the velocity at that location using the velocity model with the fitted (best-fit) parameters in table 10. Similarly, each velocity in the 20 by 20 grid as modeled by CFD was compared to the velocity generated by the velocity model with the parameters estimated by the proposed design equations. These errors are summarized in table 12 as "Fitted: Grid" and "Estimated: Grid," respectively.

Errors were also computed on the depth-averaged slices. These errors were computed using both the fitted (best-fit) parameters as well as the parameters estimated using the proposed design equations. Table 12 summarizes these results.

Table 12. Summary of RMSE estimates for the 3-ft culvert runs.

Category	Minimum (percent)	Median (percent)	Maximum (percent)
Fitted: Grid	5.3	10.1	26.4
Estimated: Grid	5.6	10.7	25.9
Fitted: Slice	2.2	3.4	6.9
Estimated: Slice	2.2	4.4	10.6

As expected, the errors using the design equations (estimated) are higher than those using the fitted parameters, but not significantly. It should be noted that for a few runs, either the grid or slice RMSE for the estimated parameters is lower than the RMSE for the fitted parameters. Recalling that the fitting used the sum of the grid and slice RMSE as the objective function, it is possible for one to decrease with the estimated parameters, but the sum of the two decreases. A complete list of RMSE results is in appendix A.

Figure 85 illustrates the slice velocities for the run with the highest fitted slice RMSE (6.9 percent). This run (C3F00V2D1) has no embedment, the higher velocity, and the shallowest depth (0.37 ft). The slice velocities predicted by the fitted and estimated parameters are nearly identical. However, they both overpredict velocity in a portion of the culvert and underpredict the depth-averaged velocity near the centerline. Inspection of the CFD data, however, suggests the mapping of the CFD data to the grid might be improved.

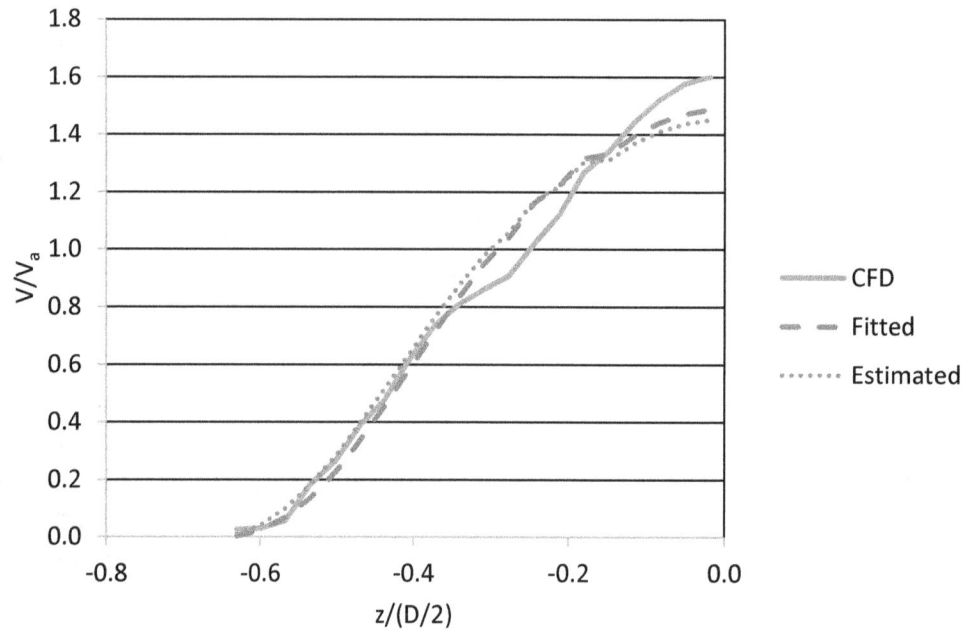

Figure 85. Graph. Comparison of velocities for C3F00V2D1.

More typical results are provided in figure 86 and figure 87, which depict runs C3F00V2D6 and C3F30V2D6, respectively. Run C3F00V2D6 has no embedment, the higher velocity, and the sixth-deepest depth (0.75 ft). Generally, the runs for the no-embedment cases have the higher RMSE values, possibly because the geometry of these runs is the full rounded culvert.

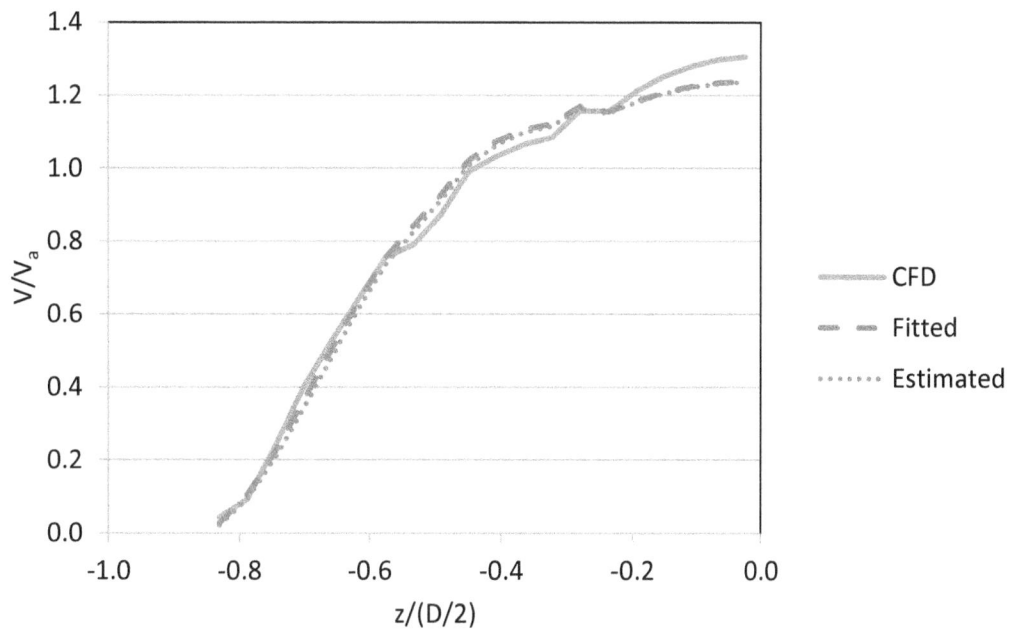

Figure 86. Graph. Comparison of velocities for C3F00V2D6.

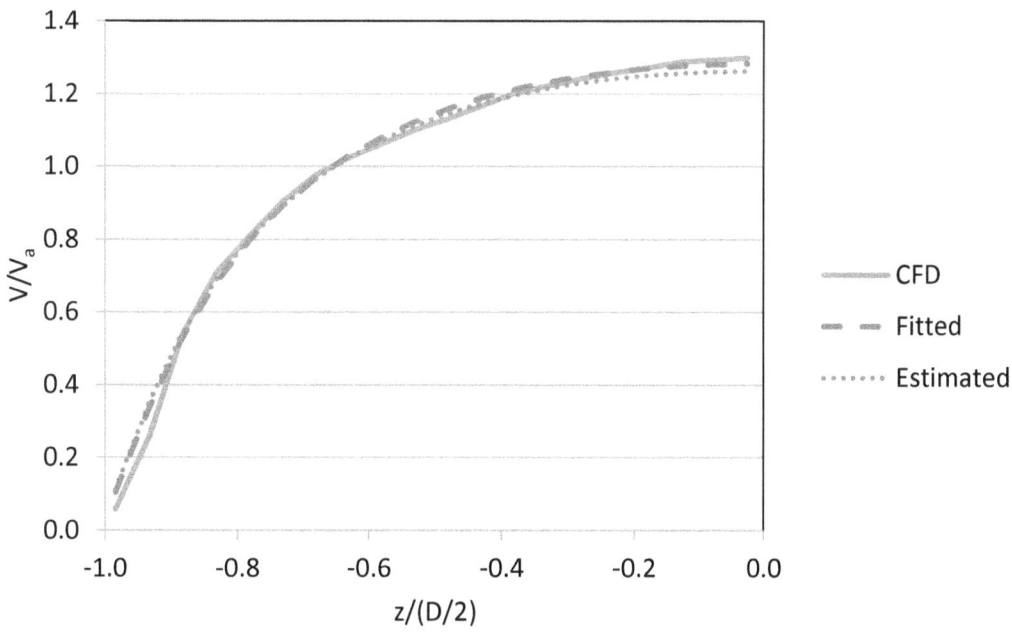

Figure 87. Graph. Comparison of velocities for C3F30V2D6.

Run C3F30V2D6 has the greatest embedment (30 percent of the culvert diameter), the higher velocity, and the sixth-deepest depth (0.75 ft). Generally, the runs with the highest embedment cases have the lowest RMSE values, possibly because the geometry of these runs is nearly rectangular. The differences in the no-embedment and 30-percent embedment cases were frequently critical to the development of the design equations.

DESIGN EQUATION VALIDATION

The design equations were developed based on the analyses of the 3-ft culvert CFD runs. These equations were validated using the 6- and 8-ft CFD runs. The 6-ft runs are scaled based on the Froude number from the 3-ft runs. The scaling was performed to test the utility of the CFD modeling but will have value in validating the design equations. The 8-ft runs are not scaled and represent a more independent set of validation data.

6-ft Culvert Validation

Fitted parameters were determined for the 6-ft culvert runs using the same approach described for the 3-ft culvert runs. These parameters are summarized in table 13. Data were not available for one run and are labeled "n/a" for "not available." Table 14 summarizes the minimum, median, and maximum RMSE for the 6-ft runs for the fitted parameters and the parameters estimated from the design equations. As expected and previously observed with the 3-ft culvert runs, the errors using the design equations (estimated) are higher than those using the fitted parameters, but not significantly. The RMSE errors for the 6-ft culvert runs are comparable to the RMSE errors for the 3-ft culvert runs, as expected. A complete listing of RMSE results for the 6-ft culvert runs is found in appendix A.

Table 13. Fitted estimates of velocity distribution parameters for the 6-ft culvert.

Run ID	M	β_i	δ_i (inches)	δ_y (inches)	ε (inches)
C6F00V1D1	-0.53	3.02	0.19	0.00	0.40
C6F00V1D2	0.27	2.97	0.63	0.00	1.85
C6F00V1D3	1.27	2.62	1.20	0.00	7.00
C6F00V2D1	-0.33	2.66	0.20	0.00	0.80
C6F00V2D2	0.24	2.90	0.88	0.00	2.89
C6F00V2D3	1.23	2.57	1.28	0.00	8.03
C6F15V1D1	1.33	1.83	0.00	0.39	0.00
C6F15V1D2	1.62	1.96	0.00	0.68	0.00
C6F15V1D3	2.49	2.15	0.00	1.28	0.00
C6F15V2D1	1.31	1.80	0.00	0.33	0.00
C6F15V2D2	1.65	1.94	0.00	0.62	0.00
C6F15V2D3	2.49	2.13	0.00	1.19	0.00
C6F30V1D1	2.29	1.35	0.00	0.03	0.00
C6F30V1D2	1.80	1.29	0.00	0.31	0.42
C6F30V1D3	2.52	1.30	0.00	0.18	0.00
C6F30V2D1	n/a	n/a	n/a	n/a	n/a
C6F30V2D2	2.53	1.36	0.00	0.16	0.00
C6F30V2D3	2.39	1.30	0.00	0.22	0.08

n/a = not available.

Table 14. Summary of RMSE estimates for the 6-ft culvert runs.

Category	Minimum (percent)	Median (percent)	Maximum (percent)
Fitted: Grid	5.7	11.7	23.5
Estimated: Grid	5.8	11.8	22.7
Fitted: Slice	2.5	3.9	6.7
Estimated: Slice	2.4	5.5	7.9

The runs shown in figure 88 and figure 89 are the scaled versions of the runs presented in the discussion of the 3-ft culvert analyses. Run C6F00V2D3 has no embedment, the higher velocity, and the deepest depth (1.5 ft) of this series. (The drop in the CFD slice velocity in the last slice, which is closest to the culvert centerline, reflects CFD output in the shallow depths that is lower in this slice than in the adjacent slice.)

Run C6F30V2D3 has the greatest embedment (30 percent of the culvert diameter), the higher velocity, and the deepest depth (1.5 ft). Generally, these runs with the highest embedment cases have the lowest RMSE values, possibly because the geometry of these runs is nearly rectangular. These results confirm the observations using the 3-ft culvert runs.

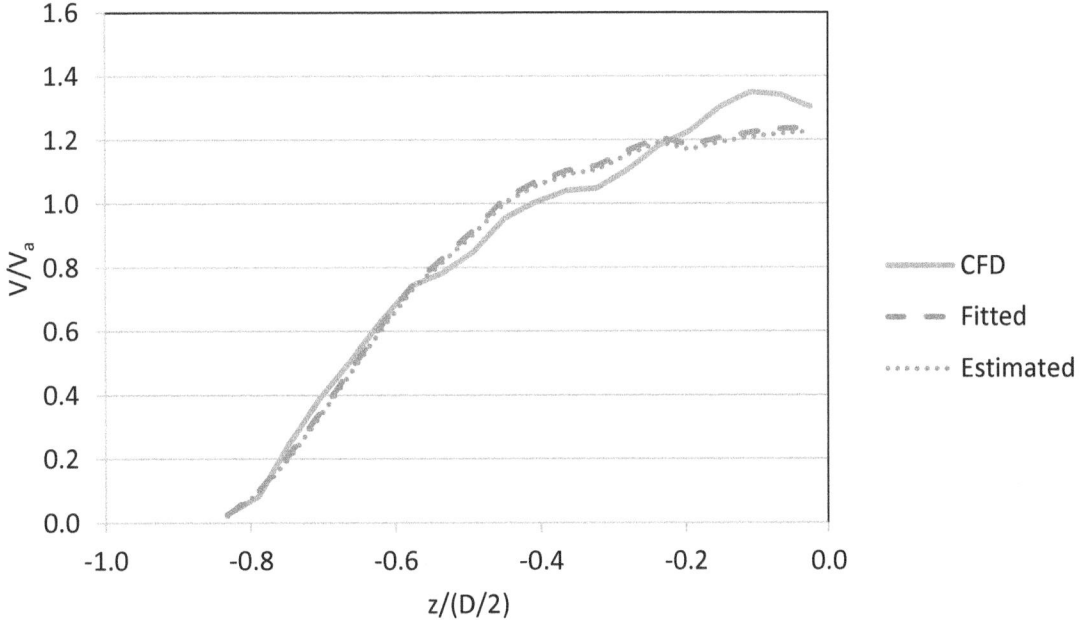

Figure 88. Graph. Comparison of velocities for C6F00V2D3.

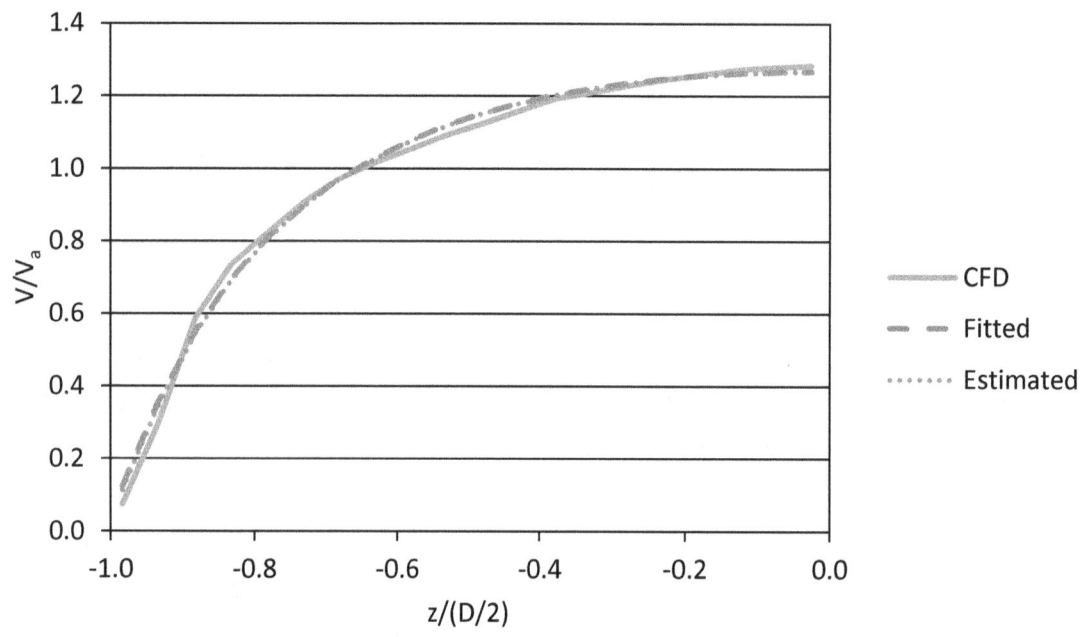

Figure 89. Graph. Comparison of velocities for C6F30V2D3.

8-ft Culvert Validation

Fitted parameters were determined for the 8-ft culvert runs using the same approach described for the 3-ft culvert runs. These parameters are summarized in table 15. Data were not available for several runs, which are labeled "n/a" for "not available." Table 16 summarizes the minimum, median, and maximum RMSE for the 8-ft runs for the fitted parameters and the parameters estimated from the design equations. As expected and previously observed with the 3-ft and 6-ft culvert runs, the errors using the design equations (estimated) are higher than those using the fitted parameters, but not significantly. Caution is appropriate in evaluating this table because it includes results from only 6 of the 18 runs in the experimental matrix. The other runs were not completed because of schedule or budget limitations. A complete listing of RMSE results, including a list of the runs completed for the 8-ft culvert runs, is found in appendix A.

The runs shown in figure 90 and figure 91 are not scaled from the 3-ft culvert but have the same relative embedment and flow depths as the runs presented in the discussions of the 3- and 6-ft culvert analyses. Run C8F00V2D3 has no embedment, the higher velocity, and the deepest depth (2.0 ft) of this series. Run C8F30V2D3 has the greatest embedment (30 percent of the culvert diameter), the higher velocity, and the deepest depth (2.0 ft). Generally, these runs with the highest embedment cases have the lowest RMSE values, possibly because the geometry of these runs is nearly rectangular. These results confirm the observations using the 3-ft and 6-ft culvert models.

Table 15. Fitted estimates of velocity distribution parameters for the 8-ft culvert.

Run ID	M	β_i	δ_i (inches)	δ_y (inches)	ε (inches)
C8F00V1D1	n/a	n/a	n/a	n/a	n/a
C8F00V1D2	n/a	n/a	n/a	n/a	n/a
C8F00V1D3	n/a	n/a	n/a	n/a	n/a
C8F00V2D1	n/a	n/a	n/a	n/a	n/a
C8F00V2D2	n/a	n/a	n/a	n/a	n/a
C8F00V2D3	1.82	2.66	1.69	0.00	15.58
C8F15V1D1	1.69	1.63	0.00	0.02	0.06
C8F15V1D2	n/a	n/a	n/a	n/a	n/a
C8F15V1D3	1.77	2.11	0.00	1.02	0.29
C8F15V2D1	n/a	n/a	n/a	n/a	n/a
C8F15V2D2	n/a	n/a	n/a	n/a	n/a
C8F15V2D3	2.28	2.06	0.00	0.50	0.00
C8F30V1D1	n/a	n/a	n/a	n/a	n/a
C8F30V1D2	n/a	n/a	n/a	n/a	n/a
C8F30V1D3	2.07	1.38	0.00	0.94	0.00
C8F30V2D1	n/a	n/a	n/a	n/a	n/a
C8F30V2D2	n/a	n/a	n/a	n/a	n/a
C8F30V2D3	2.27	1.27	0.00	0.15	0.03

n/a = not available.

Table 16. Summary of RMSE estimates for the 8-ft culvert runs.

Category	Minimum (percent)	Median (percent)	Maximum (percent)
Fitted: Grid	4.8	8.6	18.5
Estimated: Grid	5.3	10.7	17.9
Fitted: Slice	2.0	2.8	7.4
Estimated: Slice	2.8	5.0	9.5

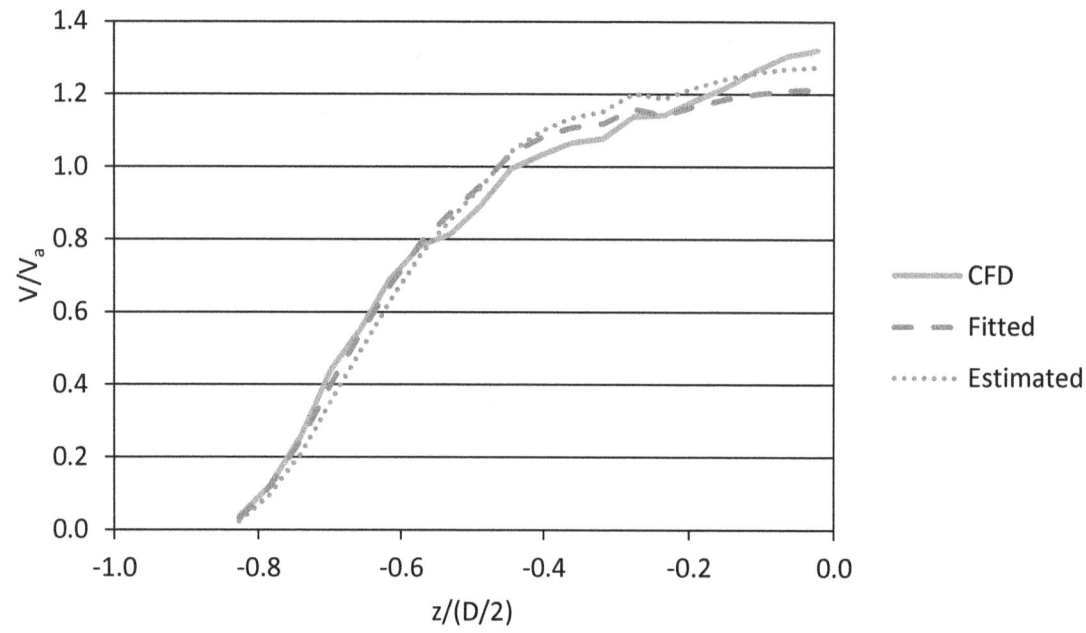

Figure 90. Graph. Comparison of velocities for C8F00V2D3.

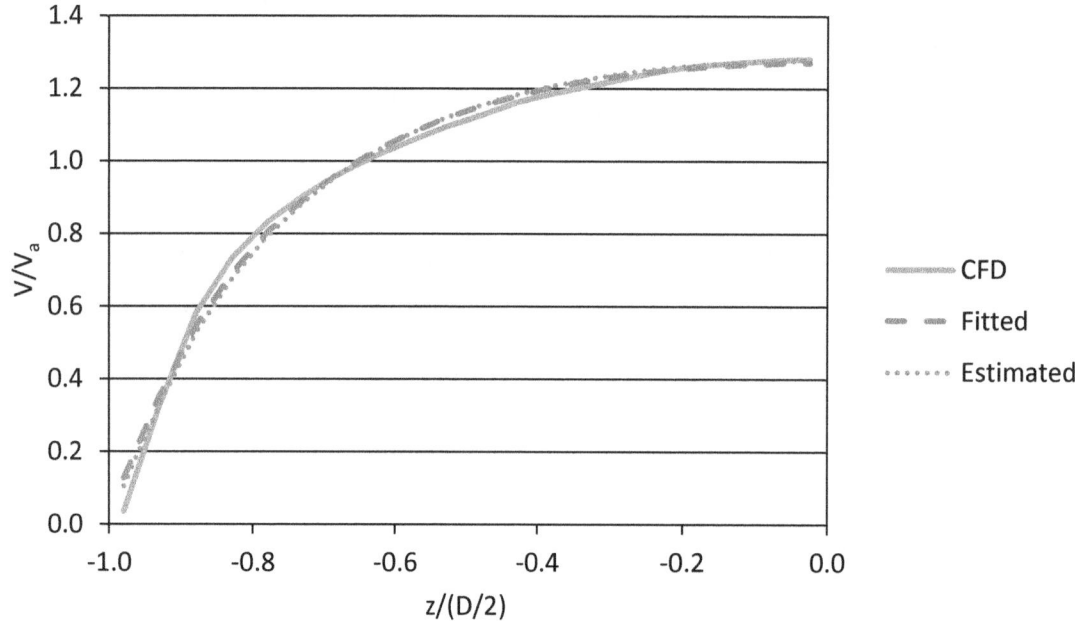

Figure 91. Graph. Comparison of velocities for C8F30V2D3.

DESIGN APPLICATION

Application of the design equations is illustrated with two examples. In both cases, the objective is to determine the portion of the culvert that can pass a particular fish requiring flow velocity less than 1.3 ft/s and flow depth of at least 0.5 ft. The design discharge in both examples is the high passage discharge, Q_H, which equals 7.8 ft³/s. The slope of the stream at the culvert crossing is 0.0015 ft/ft, which is also to be used as the slope of the culvert barrel.

76

The tailwater channel is approximated by a trapezoidal channel with a bottom width of 5.5 ft and 1:1 side slopes. The bed material has a D_{50} of 0.944 inches with a D_{84} of 2.6 inches. Assuming normal depth (0.88 ft) in the tailwater channel and using the Blodgett method, a Manning's roughness of 0.033 is estimated for the bed material.

Example 1: Embedded Culvert

In this example, a 6-ft diameter corrugated metal pipe (CMP) embedded to a depth of 2 ft is evaluated. (HEC 26 recommends minimum embedment of 30 percent of the culvert diameter or 2 ft, whichever is larger.)[1] The configuration is shown in figure 92. The Manning's roughness for the culvert is 0.028 according to HDS 5.[18] Because the embedment material will be the same material as that found in the tailwater channel, a Manning's n of 0.033 is used for the embedment.

Using a culvert analysis tool such as HY-8, the hydraulic properties within the culvert for the given conditions are determined. Normal depth is 0.91 ft, but HY-8 shows that this depth is not reached within the culvert barrel under these conditions. The depth in the barrel ranges from 0.88 ft at the culvert outlet (controlled by the tailwater depth) to a slightly deeper 0.89 ft at the culvert inlet. For consideration of fish passage, the lower depth (higher velocity) condition at the outlet is used. From the HY-8 output, the average velocity at the outlet is 1.51 ft/s. The average velocity exceeds the velocity specified for the fish of interest, suggesting this configuration is a barrier to passage.

The methods developed in this report are applied to determine whether this is true. Estimates of the following quantities and parameters are required:

1. Physical properties.

2. Velocity distribution parameters.

3. ξ_0 and ξ_{max}.

4. Overall velocity distribution (computation of a point velocity is illustrated).

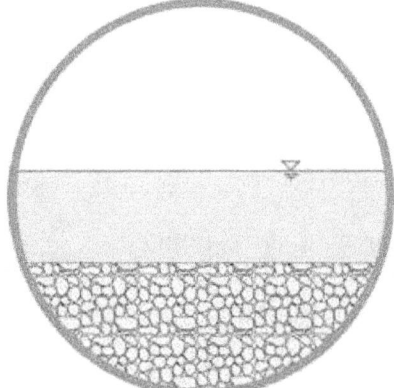

Figure 92. Illustration. Embedded culvert for example 1.

Estimate Physical Properties

To determine the velocity distribution within the culvert, the parameters M, β_i, δ_i, δ_y, and ε must be estimated. To do this, a few ratios describing the situation are required: T^2/A and B_1/B_{avg}. To develop these, one needs to compute A, T, B_1, and B_{avg} as follows:

$$A = Q/V = 7.8/1.51 = 5.17 \text{ ft}^2$$

For a circular conduit, the water surface top width T is calculated from figure 93.

$$T = D \sin(\theta)$$

Figure 93. Equation. Water surface top width.

Where:
Θ = Angle between a line to the water surface at the edge of the pipe from the center of the pipe to the vertical.

$$\theta = a\cos\left(1 - \frac{2(y_{max} + d_e)}{D}\right)$$

Figure 94. Equation. Angle to water surface.

Where:
d_e = Embedment depth, ft (m).

In this example, $y_{max} = y = 0.88$ ft; $d_e = 2.0$ ft; and $D = 6.0$ ft.

$$\Theta = a\cos(1 - 2(0.88+2.0)/6) = 87.7 \text{ degrees}$$

$$T = 6.0 \sin(87.7) = 6.0 \text{ ft}$$

For the symmetrical pipe, $B_1 = T/2 = 6.0/2 = 3.0$ ft.

$$B_{avg} = (A/2)/y_{max} = (5.17/2)/0.88 = 2.94 \text{ ft}$$

$$T^2/A = (6.0)^2/5.17 = 6.96$$

$$B_1/B_{avg} = 3.0/2.94 = 1.02$$

Estimate Velocity Distribution Parameters

Next, the velocity distribution parameters are estimated:

V_{max}/V_a is calculated from the equation in figure 73:

$$\frac{V_{max}}{V_a} = [0.047 - 0.029(1.02)][6.96]^{[-2.7+3.6(1.02)]} + 1.38 = 1.50$$

From this, $V_{max} = 1.5 \, V_a = 1.5 \, (1.51) = 2.26$ ft/s.

M is calculated from the equation in figure 70:

$$\frac{V_a}{V_{max}} = 0.667 = \frac{e^M}{e^M - 1} - \frac{1}{M}$$

By trial and error, M is calculated as 2.19.

β_i is calculated from the equation in figure 76:

$$\beta_i = 2.56\left(\frac{B_1}{B_{avg}} - 1\right)^{0.49} + 1 = 2.56(1.02 - 1)^{0.49} + 1 = 1.39$$

For this example, B_1/B_{avg} is less than 1.2. Therefore, ε and δ_i are both set to zero.

The final parameter, δ_y, is calculated using the equation in figure 84 with $a_y = 0.018$ taken from table 11:

$$\delta_y = a_y y_{max} = 0.018(0.88) = 0.016 \, ft$$

Estimate ξ_0 and ξ_{max}

The final parameters needed are the minimum and maximum isovel values. The minimum value, ξ_0, represents the lowest velocity zone, which occurs at the channel perimeter. In the design cases covered by this methodology, ξ_0 equals zero.

The maximum value, ξ_{max}, represents the highest velocity zone in the cross-section. Because $\varepsilon \geq 0$, the maximum occurs at the centerline of a symmetrical cross-section ($z = 0$) at the water surface when $y = y_{max}$, as illustrated in figure 60. The corresponding values of Y, Z, and ξ are estimated from the equations in figure 63, figure 64, and figure 62, respectively:

$$Y = \frac{y + \delta_y}{y_{max} + \delta_y + \varepsilon} = \frac{0.88 + 0.016}{0.88 + 0.016 + 0} = 1.0$$

$$Z = \frac{|z|}{B_i + \delta_i} = \frac{|0|}{3.0 + 0} = 0$$

$$\xi = Y(1 - Z)^{\beta_i} \exp(\beta_i Z - Y + 1) = 1.0(1 - 0.0)^{1.39} \exp(1.39(0) - 1.0 + 1) = 1.0$$

From the above, $\xi_{max} = 1.0$.

Example Computation of a Point Velocity

With the above parameters specific to this example, velocities throughout the flow field can be computed. The velocity at a point 0.5 ft above the bed and 1.0 ft to the left of the culvert centerline, shown in figure 95, is computed to demonstrate the process.

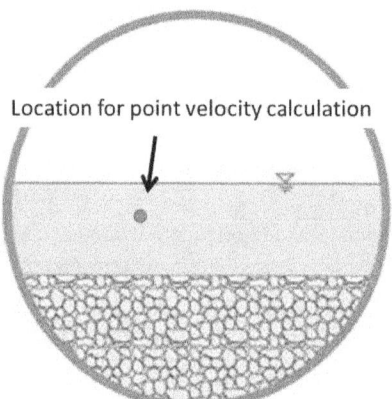

Figure 95. Illustration. Location for example 1 point velocity computation.

For this computation, $y = 0.5$ ft and $z = -1.0$ ft. The value for the isovel, ξ, is computed the same way as ξ_{max} was computed using the equations in figure 63, figure 64, and figure 62, respectively:

$$Y = \frac{y + \delta_y}{y_{max} + \delta_y + \varepsilon} = \frac{0.5 + 0.016}{0.88 + 0.016 + 0} = 0.576$$

$$Z = \frac{|z|}{B_i + \delta_i} = \frac{|-1|}{3.0 + 0} = 0.333$$

$$\xi = Y(1 - Z)^{\beta_i} \, exp(\beta_i Z - Y + 1) = 0.576(1 - 0.333)^{1.39} \, exp(1.39(0.333) - 0.576 + 1) = 0.796$$

With this information, one calculates the point velocity from the equation in figure 65:

$$V = \frac{1}{M} ln\left[1 + \left(e^M - 1\right)\frac{\xi - \xi_0}{\xi_{max} - \xi_0} \right]V_{max} = \frac{1}{2.19} ln\left[1 + \left(e^{2.19} - 1\right)\frac{0.796 - 0}{1 - 0} \right]2.26 = 2.05 \, ft/s$$

Therefore, the estimated velocity at a point in the flow field 0.5 ft above the bed and 1.0 ft to the left of the culvert centerline is 2.05 ft/s.

Estimating the Overall Velocity Distribution

To evaluate the entire cross-section, point velocities may be computed on a rectangular grid, such as the 20 by 20 grid used in previous analyses. Slice velocities can then be computed by depth-averaging the grid point values. The result of this process for example 1 is summarized in table 17. The slices are numbered from the left edge of the flow cross-section looking downstream through the culvert. Higher velocities are observed moving to the right toward the culvert centerline. Because this cross-section approximates a rectangular shape, the depth after slice 1 is constant at 0.88 ft. Velocity and depth are shown graphically in figure 96.

For the design criteria (velocity ≤ 1.3 ft/s and depth ≥ 0.5 ft), this design satisfies the depth requirement across the entire width of the flow cross-section except a narrow part of slice 1. However, the velocity criterion is met for only the six leftmost slices because the velocity in these slices is less than 1.3 ft/s. Passage is facilitated when both criteria are satisfied. For this case, both criteria are satisfied in the six leftmost slices. Because each slice represents a width of 0.15 ft (B_1 divided by 20), 0.9 ft of flow width (6 times 0.15) meets the design criteria for passage, allowing for passage to occur even though the average velocity suggests that would not be the case. Because the culvert and flow field are symmetrical, a mirror image passage path also exists on the right side.

Table 17. Slice velocity and depths for example 1.

Slice Number	Distance from Centerline (ft)	Velocity (ft/s)	Depth (ft)
1	-2.92	0.16	0.57
2	-2.77	0.46	0.88
3	-2.62	0.74	0.88
4	-2.47	0.96	0.88
5	-2.32	1.14	0.88
6	-2.17	1.28	0.88
7	-2.02	1.40	0.88
8	-1.87	1.50	0.88
9	-1.72	1.58	0.88
10	-1.57	1.65	0.88
11	-1.42	1.70	0.88
12	-1.27	1.75	0.88
13	-1.12	1.79	0.88
14	-0.97	1.82	0.88
15	-0.82	1.84	0.88
16	-0.67	1.86	0.88
17	-0.52	1.88	0.88
18	-0.37	1.89	0.88
19	-0.22	1.90	0.88
20	-0.07	1.90	0.88

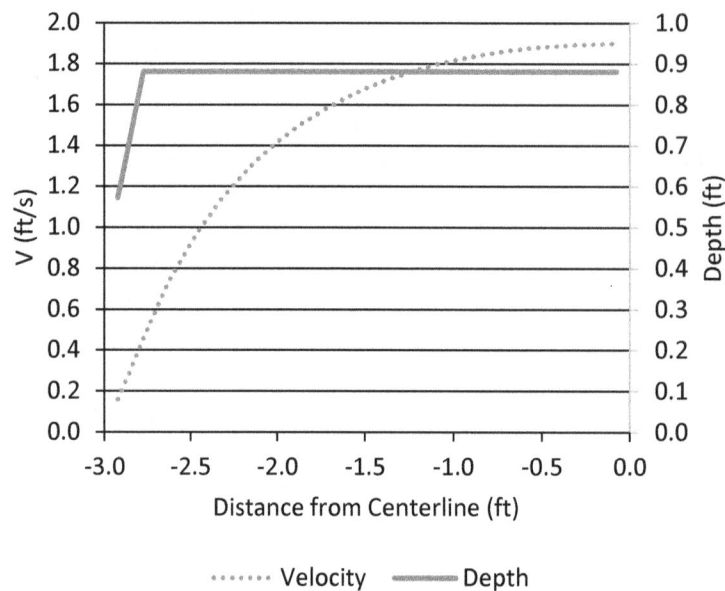

Figure 96. Graph. Velocity and depth for example 1.

Example 2: Non-embedded Culvert

In this example, a 6-ft diameter CMP without embedment is evaluated. The configuration is shown in figure 97. There is no bed material within the culvert, and the Manning's roughness for the culvert is 0.028 according to HDS 5.[18]

Using a culvert analysis tool, such as HY-8, the hydraulic properties within the culvert for the given conditions are determined. Normal depth is 1.27 ft, but HY-8 shows that this depth is not reached within the culvert barrel under these conditions. The depth in the barrel ranges from 0.88 ft at the culvert outlet (controlled by the tailwater depth) to 1.14 ft (approaching normal) at the culvert inlet. For consideration of fish passage, the lower depth (higher velocity) condition at the outlet is used. From the HY-8 output, the average velocity is 3.01 ft/s at the culvert outlet. The average velocity exceeds the velocity specified for the fish of interest, suggesting this configuration is a barrier to passage.

The methods developed in this report are applied to determine whether this is true. Estimates of the following quantities and parameters are required:

1. Physical properties.

2. Velocity distribution parameters.

3. ξ_0 and ξ_{max}.

4. Overall velocity distribution (computation of a point velocity is illustrated).

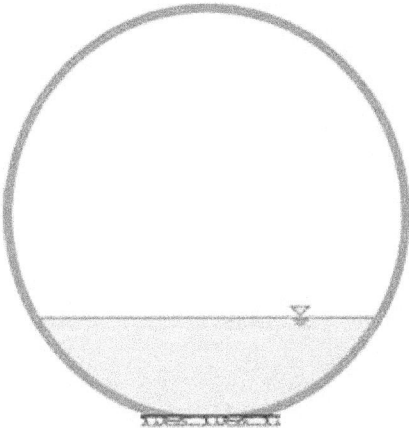

Figure 97. Illustration. Culvert without embedment for example 2.

Estimate Physical Properties

To determine the velocity distribution within the culvert, the parameters M, β_i, δ_i, δ_y, and ε must be estimated. To do this, a few ratios describing the situation are required: T^2/A and B_1/B_{avg}. To develop these one needs to compute A, T, B_1, and B_{avg} as follows:

$$A = Q/V = 7.8/3.01 = 2.59 \text{ ft}^2$$

For a circular conduit, the water surface top width T is calculated from the equations in figure 93 and figure 94. In this example, $y_{max} = y = 0.88$ ft; $d_e = 0.0$ ft; and $D = 6.0$ ft.

$$\Theta = \text{acos } (1 - 2(0.88+0.0)/6) = 45.0 \text{ degrees}$$

$$T = 6.0 \sin(45.0) = 4.24 \text{ ft}$$

For the symmetrical pipe, $B_1 = T/2 = 4.24/2 = 2.12$ ft

$$B_{avg} = (A/2)/y_{max} = (2.59/2)/0.88 = 1.47 \text{ ft}$$

$$T^2/A = (4.24)^2/2.59 = 6.94$$

$$B_1/B_{avg} = 2.12/1.47 = 1.44$$

Estimate Velocity Distribution Parameters

Next, the velocity distribution parameters are estimated:

V_{max}/V_a is calculated from the equation in figure 73:

$$\frac{V_{max}}{V_a} = [0.047 - 0.029(1.44)][6.94]^{[-2.7+3.6(1.44)]} + 1.38 = 2.02$$

From this, $V_{max} = 2.02\ V_a = 2.02\ (3.01) = 6.08$ ft/s.

M is calculated from the equation in figure 70:

$$\frac{V_a}{V_{max}} = 0.495 = \frac{e^M}{e^M - 1} - \frac{1}{M}$$

By trial and error, M is calculated as -0.12.

B_i is calculated from the equation in figure 76:

$$\beta_i = 2.56\left(\frac{B_1}{B_{avg}} - 1\right)^{0.49} + 1 = 2.56(1.44 - 1)^{0.49} + 1 = 2.71$$

For this example, B_1/B_{avg} is greater than or equal to 1.2. Therefore, ε is nonzero. To estimate ε, the Froude number based on average cross-section velocity and depth ($y_a = A/T$) is required:

$$Fr = \frac{V_a}{\sqrt{gy_a}} = \frac{3.01}{\sqrt{32.2(2.59/4.24)}} = 0.68$$

Using the equation in figure 81, ε is calculated as follows:

$$\varepsilon = 2.56\left(\frac{T^2}{A}\right)^{-1.54} (Fr)^{-0.62} y_{max} = 2.56(6.94)^{-1.54}(0.68)^{-0.62}0.88 = 0.14$$

Because B_1/B_{avg} is greater than or equal to 1.2, the equation in figure 83 is used to calculate δ_i as follows:

$$\delta_i = 0.04B_1 = 0.04(2.12) = 0.085$$

Because B_1/B_{avg} is greater than or equal to 1.2, the final parameter, δ_y, is set equal to zero.

Estimate ξ_0 and ξ_{max}

The final parameters needed are the minimum and maximum isovel values. The minimum value, ξ_0, represents the lowest velocity zone, which occurs at the channel perimeter. In the design cases covered by this methodology, ξ_0 equals zero.

The maximum value, ξ_{max}, represents the highest velocity zone in the cross-section. Because $\varepsilon \geq 0$, the maximum occurs at the centerline of a symmetrical cross-section ($z = 0$) at the water surface when $y = y_{max}$ as illustrated in figure 60. The corresponding values of Y, Z, and ξ are estimated from the equations in figure 63, figure 64, and figure 62, respectively.

$$Y = \frac{y + \delta_y}{y_{max} + \delta_y + \varepsilon} = \frac{0.88 + 0.0}{0.88 + 0.0 + 0.14} = 0.863$$

$$Z = \frac{|z|}{B_i + \delta_i} = \frac{|0|}{2.12 + 0.085} = 0$$

$$\xi = Y(1-Z)^{\beta_i} \, exp(\beta_i Z - Y + 1) = 0.863(1-0.0)^{2.71} \, exp(2.71(0) - 0.863 + 1) = 0.99$$

From the above, $\xi_{max} = 0.99$.

Example Computation of a Point Velocity

With the above parameters specific to this example, velocities throughout the flow field can be computed. The velocity at a point 0.5 ft above the bed and 1.0 ft to the left of the culvert centerline, shown in figure 98, is computed to demonstrate the process.

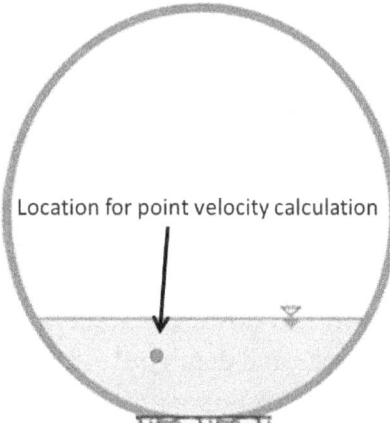

Figure 98. Illustration. Location for example 2 point velocity computation.

For this computation, $y = 0.5$ ft and $z = -1.0$ ft. The value for the isovel, ξ, is computed the same way as ξ_{max} was computed using the equations in figure 63, figure 64, and figure 62, respectively:

$$Y = \frac{y + \delta_y}{y_{max} + \delta_y + \varepsilon} = \frac{0.5 + 0.0}{0.88 + 0.0 + 0.14} = 0.490$$

$$Z = \frac{|z|}{B_i + \delta_i} = \frac{|-1|}{2.12 + 0.085} = 0.453$$

$$\xi = Y(1 - Z)^{\beta_i} \, exp(\beta_i Z - Y + 1) = 0.490(1 - 0.453)^{2.71} \, exp(2.71(0.453) - 0.490 + 1) = 0.543$$

With this information, one calculates the point velocity from the equation in figure 65:

$$V = \frac{1}{M} ln\left[1 + \left(e^M - 1\right)\frac{\xi - \xi_0}{\xi_{max} - \xi_0}\right]V_{max} = \frac{1}{-0.12} ln\left[1 + \left(e^{-0.12} - 1\right)\frac{0.543 - 0}{0.99 - 0}\right]6.08 = 3.2 \, ft/s$$

Therefore, the estimated velocity at a point in the flow field 0.5 ft above the bed and 1.0 ft to the left of the culvert centerline is 3.2 ft/s.

Estimating the Overall Velocity Distribution

To evaluate the entire cross-section, point velocities may be computed on a rectangular grid, such as the 20 by 20 grid used in previous analyses. Slice velocities can then be computed by depth-averaging the grid point values. The result of this process for example 2 is summarized in table 18. The slices are numbered from the left edge of the flow cross-section looking downstream through the culvert. Higher velocities are observed moving to the right toward the culvert centerline. Depth is shown graphically in figure 99, and velocity is shown in figure 100.

For the design criteria (velocity ≤ 1.3 ft/s and depth ≥ 0.5 ft), this design does not meet the depth requirement in the first six slices of the flow cross-section. However, the depth is sufficient for passage to the right of this point as shown in figure 99. The velocity criterion is met only for the five leftmost slices, as seen in figure 100. For passage to occur, both criteria must be satisfied in the same area of the culvert. Therefore, in this case there is no part of the flow field that satisfies both the depth and velocity criteria; the flow velocity is too high, the flow depth is too low, or both (as in slice 6). Therefore, this culvert is a fish passage barrier.

Table 18. Slice velocity and depths for example 2.

Slice Number	Distance from Centerline (ft)	Velocity (ft/s)	Depth (ft)
1	-2.07	0.04	0.04
2	-1.96	0.16	0.13
3	-1.86	0.37	0.22
4	-1.75	0.66	0.31
5	-1.65	1.00	0.40
6	-1.54	1.39	0.44
7	-1.43	1.76	0.53
8	-1.33	2.16	0.57
9	-1.22	2.53	0.62
10	-1.11	2.87	0.66
11	-1.01	3.16	0.70
12	-0.90	3.39	0.75
13	-0.80	3.55	0.79
14	-0.69	3.81	0.79
15	-0.58	3.86	0.84
16	-0.48	4.03	0.84
17	-0.37	4.16	0.84
18	-0.27	4.07	0.88
19	-0.16	4.13	0.88
20	-0.05	4.15	0.88

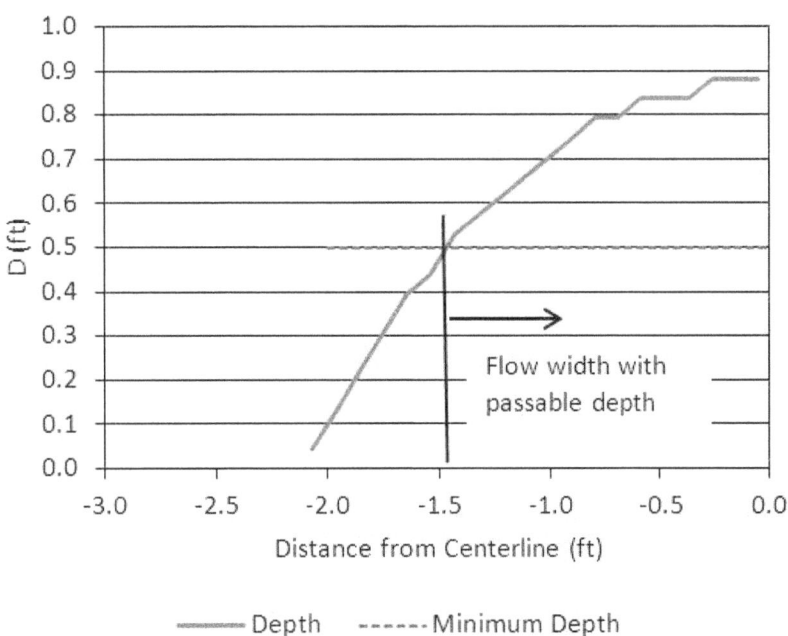

Figure 99. Graph. Depth for example 2.

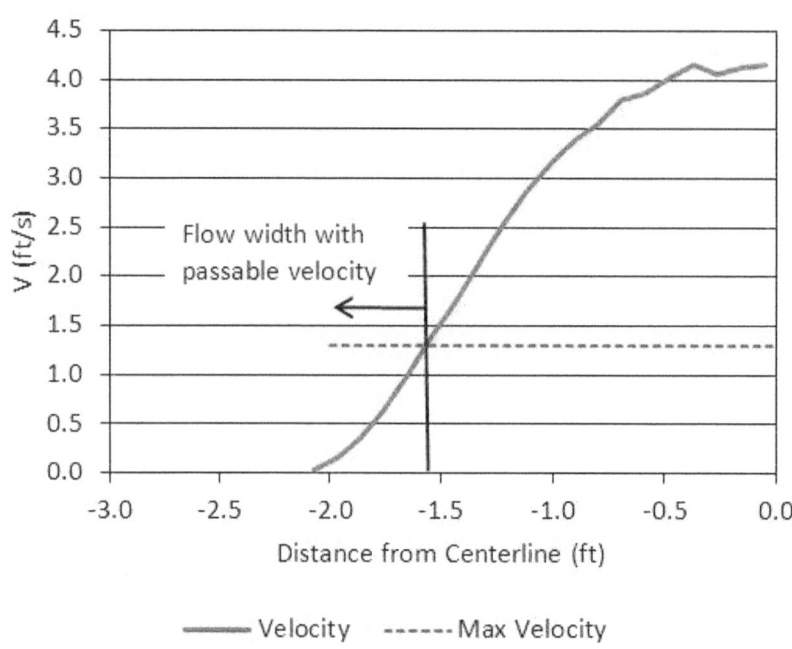

Figure 100. Graph. Velocity for example 2.

Observations

Both design examples show cross-section average velocities in excess of the maximum required velocity for the target fish. Using the same culvert size in both cases, the case without embedment (example 2) provides no locations where both depth and velocity criteria are met. However, using the same culvert with 2.0 ft of embedment, a passable width in the culvert is provided. It is important to note that this culvert is sunk 2.0 ft in example 1 compared with example 2.

The designer should repeat these analyses at the low passage flow, Q_L. The high and low passage flows bracket when passage is likely to be required. These flows may be specified by the appropriate regulatory agency to specifically correspond with the fish species of interest. The designer should also verify that the embedded culvert retains sufficient capacity to satisfy the applicable peak flow requirements.

It should be noted that the combination of the depth and embedment in example 1 forces the water surface above the culvert spring line, where further increases in depth will result in a decrease in top width, T. The design equations were not developed for relative depths above those included in this study and should be used with caution for depths above the spring line. However, because the decrease in width is within the 2-percent guideline provided previously, the use of these equations is valid in this example.

In addition, this methodology focuses on the main culvert barrel, not the inlet and outlet where flow transitions will modify the velocity profile. Other complicating factors could include the presence of debris and nonuniform bed conditions within the culvert.

VELOCITY DESIGN CHART DEVELOPMENT (CHANG'S METHOD)

The design equations and applications previously demonstrated can be applied using computer aided tools, for example, spreadsheets or custom software. When such tools are unavailable, the simplified design charts provided in this section may be used. There will be some loss of accuracy, but application is accomplished without advanced tools.

The design charts are derived from the concepts that Dr. Fred Chang developed when this research project was being developed. Chang hypothesized that a generalized series of curves describing the ratio of the cumulative cross-section average velocity from the left edge of the culvert to the cross-section average velocity could be developed. A conceptual example is shown in figure 101.

Chang suggested that such a tool could be used to estimate the fish path width, given the average culvert velocity and the fish passage velocity. For example, if the average culvert velocity is 1.5 ft/s and the maximum fish passage velocity is 1.3 ft/s, then such a tool can be used to find the portion of the culvert cross-section that averages 1.3 ft/s. To do so, the designer would take the ratio of $1.3/1.5 = 0.87$ and enter the design chart on the ordinate axis at this value. One would read an abscissa value of approximately -0.3. This means that the average cross-sectional velocity from the left edge of the water surface in the culvert to this point is 1.3 ft/s. The fish passage width, therefore, is the leftmost 70 percent of the flow width. If the culvert is symmetrical at its center, then there would be a mirror image fish passage path on the right side of the culvert.

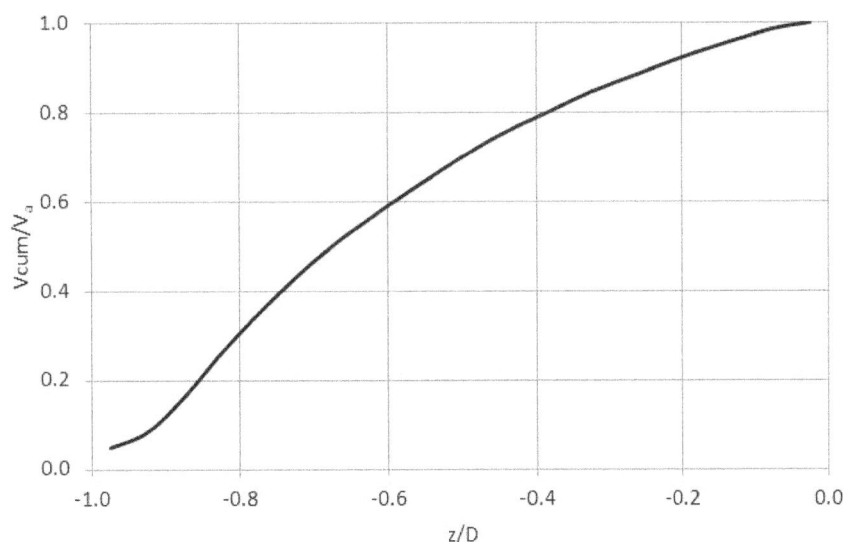

Figure 101. Graph. Conceptual cumulative velocity relation.

However, this approach overestimates the fish passage width because the average velocity includes vertical slices that are less than the average (toward the left edge) and vertical slices that are greater than the average (toward the culvert center). Therefore, this entire width is not available for fish passage.

This problem is addressed by considering the vertically averaged velocities directly rather than cumulative averages as described in the previous section of this report. (This approach will still produce some vertical slices where the velocities near the surface exceed the fish passage velocity, but this occurrence is reduced compared with the cumulative approach discussed in this section.)

Given this limitation, the data available from this study were analyzed, and it was determined that four design curves could reasonably represent the range of circular culvert conditions studied. These four design curves are summarized as follows:

1. Culvert with no embedment and T/D is less than 0.8.

2. Culvert with no embedment and T/D is greater than or equal to 0.8.

3. Culvert with embedment less than 20 percent of the culvert diameter. (Note that HEC 26 recommends a minimum embedment of 30 percent for circular culverts.[1])

4. Culvert with embedment greater than or equal to 20 percent of the culvert diameter.

These curves are shown graphically in figure 102 and in table 19. These curves were selected after examining the CFD runs for differences based on Froude numbers, Reynolds numbers, and the geometrical ratios T/D and B_1/B_{avg}. Although the CFD data included culvert diameters between 3 and 8 ft, inclusively, the dimensionless curves should be applicable to larger culverts.

No data were developed in this study for box shapes. However, design curve 4 may be used for boxes because the flow cross-section shapes for higher levels of embedment closely approximate a rectangular shape. This curve applies for all levels of embedment in box shapes.

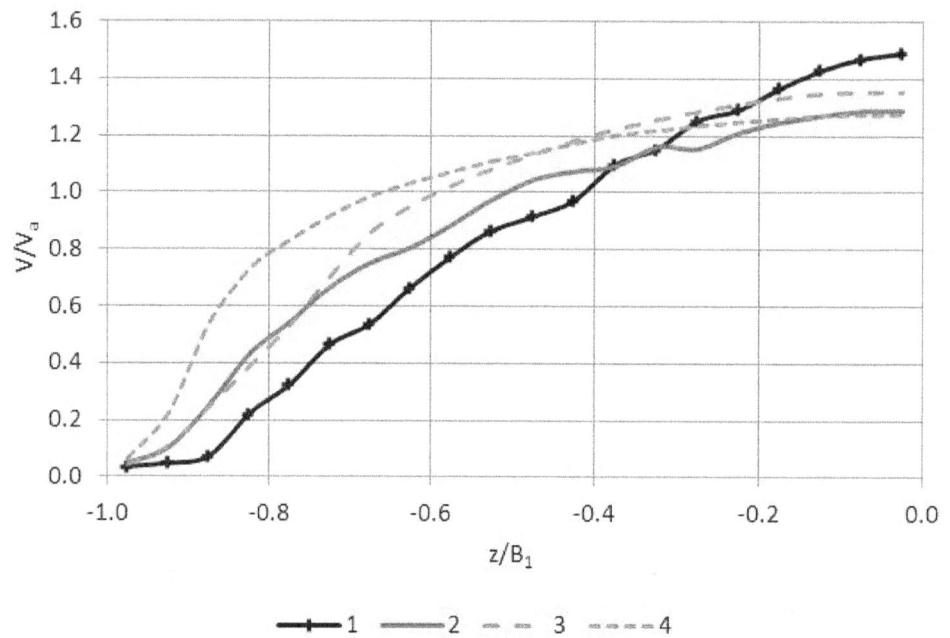

Figure 102. Graph. Vertical slice design curves.

Using the values from the previous illustration (average culvert velocity = 1.5 ft/s and the maximum fish passage velocity = 1.3 ft/s), one can find the portion of the culvert cross-section that has vertical slices with velocities less than or equal to 1.3 ft/s. With this approach, one would also have to know more about the site condition. It is assumed that this is a culvert with 30-percent embedment; therefore, design curve 4 is the appropriate choice. By inspection of either table, the ratio of 0.87 (1.3/1.5) is exceeded in vertical slice 6. That means that the first five slices provide velocities less than fish passage velocity. Because each slice represents one-twentieth of one half of the water surface flow width, the fish passage width is the leftmost 25 percent (5 times 1/20) of the flow width from the left edge to the culvert centerline. The same result is obtained by using the figure. Use of these design curves is also illustrated in the application guide provided in appendix C of this document.

Table 19. Tabular listing of vertical slice design curves.

Slice	z/B_1	V/V_a Curve 1	V/V_a Curve 2	V/V_a Curve 3	V/V_a Curve 4
1	-0.975	0.036	0.048	0.043	0.064
2	-0.925	0.047	0.103	0.108	0.225
3	-0.875	0.072	0.252	0.246	0.537
4	-0.825	0.221	0.430	0.386	0.729
5	-0.775	0.322	0.542	0.529	0.830
6	-0.725	0.465	0.667	0.705	0.917
7	-0.675	0.537	0.751	0.857	0.985
8	-0.625	0.666	0.806	0.951	1.032
9	-0.575	0.774	0.884	1.021	1.071
10	-0.525	0.865	0.974	1.081	1.108
11	-0.475	0.915	1.044	1.134	1.140
12	-0.425	0.972	1.073	1.182	1.172
13	-0.375	1.097	1.092	1.222	1.201
14	-0.325	1.153	1.160	1.256	1.218
15	-0.275	1.253	1.155	1.285	1.235
16	-0.225	1.291	1.211	1.310	1.248
17	-0.175	1.366	1.246	1.332	1.259
18	-0.125	1.428	1.269	1.348	1.270
19	-0.075	1.470	1.287	1.353	1.274
20	-0.025	1.488	1.287	1.353	1.274

APPLICATION EXTENSIONS

The analyses and tools from this study are based on round corrugated metal culverts with and without embedment. However, extension of these tools to other situations is possible.

Pipe Arch, Elliptical, and Box Shapes

As described in HDS 5, metal culverts are also available in pipe arch, elliptical, and box shapes, which also have the potential to serve as embedded culverts.[18] As was observed in the analyses of the embedded circular culverts, the flow area approximates a rectangle in many of the cases tested. Therefore, it is reasonable to assume that these methods are applicable to box shapes. These methods may also apply to elliptical culverts.

However, pipe arch shapes have the characteristic that at very low combinations of embedment and flow depth, the top width of the flow field decreases with increasing water surface elevation. Such an occurrence is outside the range of most tests and not well suited for the velocity distribution model that assumes increasing top width with increasing depth. For those conditions in a pipe arch where this is true, the tools developed in this study are considered applicable. For conditions beyond this, the method is not recommended.

Concrete Culverts

The design methods developed for this study rely primarily on the average velocity and cross-section geometry of the flow field. Nothing in the study, particularly the variation in bed material roughness, indicates that the methods are restricted to metal culverts. In addition, analyses were provided for computing a composite roughness for an embedded culvert that are equally applicable to concrete embedded culverts. Therefore, the methods developed for this study are anticipated to be applicable to concrete culverts.

Ledges for Terrestrial Organisms

A series of CFD runs were conducted to evaluate potential effects on the velocity distribution and composite roughness of various ledge configurations intended to provide a dry passage through the culvert for terrestrial organisms. Two alternative configurations were considered—a shelf ledge and a bench ledge—as illustrated in figure 103 and figure 104, respectively. In both cases, the design objective is to provide passage for terrestrial organisms simultaneously with fish passage. Therefore, the top of the ledge should be at or above the water surface elevation for the high passage flow.

For the shelf ledge, the shelf will have little or no influence on the velocity distribution in the flow field, depending on the means of support for the shelf. At flood flows, where the ledge is submerged, the designer should consider the loss of conveyance and increased drag resulting from the presence of the shelf.

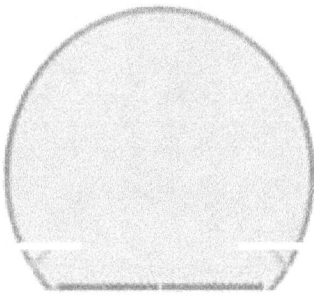

Figure 103. Illustration. Shelf ledge configuration.

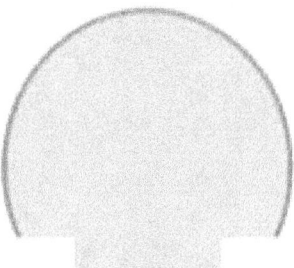

Figure 104. Illustration. Bench ledge configuration.

For the bench ledge, the bench eliminates conveyance for both passage and flood flows. As can be seen in figure 104, the remaining cross-sectional area for flow rates with a water surface elevation less than or equal to the top elevation of the bench (fish passage flows) is a rectangular area. Therefore, the methods in this report can be used to evaluate the velocity distribution. As with the shelf ledge, during flood flows when the ledge is submerged, the designer should consider the loss of conveyance and increased drag resulting from the presence of the bench.

Manning's n computations were made for the ledge configurations within an 8-ft CSP pipe with 15-percent embedment. The bed material was represented by a single layer of spheres with a diameter of 2.28 inches. The flow conditions were for full flow with a velocity of 1 ft/s.

For the shelf ledge configuration shown in figure 103, the composite roughness was estimated at 0.028. For the same conditions without the ledge, the composite roughness was estimated at 0.030. An increase is counterintuitive because the presence of the ledge should increase roughness. One possible explanation is that the ledge is represented in the CFD model with no roughness.

For the bench ledge configuration shown in figure 104, the composite roughness was estimated at 0.029. In this case, the bench surfaces are also modeled without roughness, so it is expected that the results would be similar to the shelf ledge. It should be noted that this configuration has a lower cross-sectional area. Because the CFD runs were performed with constant velocity, the bench ledge configuration run represents a lower discharge than was analyzed for the shelf ledge.

Results for both ledge configurations are on the low side of expectations for an 8-ft CSP with a gravel bed material and a ledge. Previous CFD runs for the 8-ft CSP flowing full without embedment resulted in a Manning's n of 0.0325. Therefore, the roughness values for the ledge runs are considered inconclusive.

CHAPTER 7. CONCLUSIONS AND RECOMMENDATIONS

This report documents a series of physical and numerical modeling efforts designed to support the development of a design procedure for characterizing the variation in velocity within non-embedded and embedded culverts. Physical modeling of symmetrical half-section circular culverts was conducted to provide data with which CFD modeling could be validated. Use of CFD allows more extensive test matrices to be evaluated in less time and budget than physical modeling.

The initial CFD modeling featured two-phase numerical computations that successfully reproduced the physical modeling results. To simplify the CFD modeling, single-phase modeling and truncated single-phase modeling were evaluated with good results. For the embedded culvert runs, a successful strategy for representing natural bed material within the culvert was developed.

Once the CFD modeling of the physical data was completed and validated, the CFD modeling was used to analyze the full culvert cross-sections (3-ft CMP) rather than symmetrical half sections. Subsequent test matrices included CFD runs scaled up to larger culvert sizes. One series of runs maintained Froude number-based scaling (6-ft CMP), and one series tested larger sizes without the scaling constraint (8-ft CMP). These full culvert cross-sections were used to develop and test the proposed design method.

The velocity distribution model that formed the basis of the design methodology is described by Chiu and Chiou and then Chiu.[30,31] Using the 42 CFD runs for a 3-ft diameter culvert, the five parameters necessary for the velocity model (β_i, δ_y, ε, M, δ_i) were estimated. Based on geometric and hydraulic parameters available to a designer, relations were developed to estimate those parameters for design purposes. The approach was successfully validated on CFD runs for 6-ft and 8-ft diameter culvert models.

The proposed design procedure allows a designer to estimate the velocity throughout a cross-section. These data may be depth-averaged to provide a distribution of velocity and depth across the culvert cross-section that may be used to evaluate fish passage. Although developed for circular culverts, the parameters used in the method are such that the procedure should be applicable to rectangular and other shapes. Two design examples are provided to illustrate the method and the required computations.

Future work is recommended to further validate and improve the recommended design methodology. Further CFD runs should capture the following qualities:

- A wider range of Froude numbers. In this study, Froude numbers (based on average velocity and average depth) ranged from 0.13 to 0.44.

- More variation in bed roughness values, both in an absolute sense and in a relative sense. In this study, only one bed roughness value was tested in detail.

- Shapes other than circular. Rectangular shapes will mimic the results of the embedded circular shapes, so these are not a priority. Testing of elliptical, pipe arch, and other shapes may be useful.

- Water depth/embedment scenarios where the water surface elevation is significantly higher than the elevation at maximum culvert width, resulting in decreasing top widths with increases in water surface elevation. The recommendation from this study is that the top width, T, should not decrease more than 2 percent from the maximum value of top width when applying the methods from this study.

APPENDIX A. DATA SUMMARY

The following tables summarize CFD velocity distributions compared with distributions computed from fitted and estimated parameters for the 3-, 6-, and 8-ft culvert runs. One 6-ft and several 8-ft runs were not completed because of schedule or budget reasons. These are indicated by "n/a" (not available) in the tables.

Table 20. 3-ft culvert run data.

Run ID	V_{max} (ft/s)	Fitted RMSE (percent)		Estimated RMSE (percent)	
		Grid	Slice	Grid	Slice
C3F00V1D1	1.53	26.4	5.0	25.9	4.5
C3F00V1D2	1.43	25.5	4.0	25.0	3.9
C3F00V1D3	1.32	23.9	2.8	23.7	3.3
C3F00V1D4	1.25	22.8	2.6	22.7	2.8
C3F00V1D5	1.21	20.8	3.3	20.9	4.0
C3F00V1D6	1.17	21.0	3.7	21.4	4.6
C3F00V1D7	1.14	21.1	3.7	21.4	4.3
C3F00V2D1	2.45	25.6	6.9	25.3	8.2
C3F00V2D2	2.21	25.0	3.5	24.4	3.4
C3F00V2D3	2.05	23.2	2.2	22.6	2.2
C3F00V2D4	1.96	21.8	2.6	21.7	2.7
C3F00V2D5	1.88	20.0	3.3	20.1	3.3
C3F00V2D6	1.82	20.4	3.7	20.6	3.8
C3F00V2D7	1.77	20.6	3.8	20.9	3.9
C3F15V1D1	1.17	10.1	5.5	9.9	5.5
C3F15V1D2	1.24	23.7	3.7	24.6	10.6
C3F15V1D3	1.12	10.0	4.3	10.6	6.0
C3F15V1D4	1.08	10.1	3.7	10.7	5.5
C3F15V1D5	1.04	11.1	4.2	10.9	4.5
C3F15V1D6	1.04	10.8	3.1	11.6	5.2
C3F15V1D7	1.02	11.7	3.1	12.0	4.3
C3F15V2D1	1.75	8.1	5.1	8.7	5.2
C3F15V2D2	1.78	8.6	4.1	9.3	5.4
C3F15V2D3	1.73	8.6	3.6	9.3	5.4
C3F15V2D4	1.69	9.0	2.9	9.6	4.9
C3F15V2D5	1.63	10.4	3.2	10.9	5.2
C3F15V2D6	1.63	9.7	2.2	10.6	4.4
C3F15V2D7	1.58	10.7	2.6	11.2	4.5
C3F30V1D1	1.05	6.3	4.7	8.2	7.0
C3F30V1D2	1.06	6.2	4.3	7.4	5.9
C3F30V1D3	1.06	6.6	3.9	6.9	4.6

		Fitted RMSE (percent)		Estimated RMSE (percent)	
C3F30V1D4	1.03	6.9	3.0	7.2	3.5
C3F30V1D5	1.04	6.8	3.0	6.7	3.2
C3F30V1D6	1.03	6.4	2.9	6.5	2.9
C3F30V1D7	1.02	6.4	2.8	6.7	3.1
C3F30V2D1	1.64	5.3	4.2	6.8	5.8
C3F30V2D2	1.66	5.4	3.8	6.0	4.5
C3F30V2D3	1.67	6.0	3.3	5.6	3.4
C3F30V2D4	1.64	6.5	2.6	6.1	2.6
C3F30V2D5	1.64	6.5	2.5	5.9	2.5
C3F30V2D6	1.62	6.1	2.5	5.9	2.9
C3F30V2D7	1.60	6.0	2.5	6.5	3.9

Table 21. 6-ft culvert run data.

Run ID	V_{max} (ft/s)	Fitted RMSE (percent)		Estimated RMSE (percent)	
		Grid	Slice	Grid	Slice
C6F00V1D1	2.19	22.5	4.5	22.4	7.9
C6F00V1D2	1.91	20.8	3.1	20.7	6.0
C6F00V1D3	1.66	20.2	5.3	20.5	6.3
C6F00V2D1	3.29	23.5	6.7	22.7	7.7
C6F00V2D2	2.99	19.3	5.7	19.2	6.7
C6F00V2D3	2.59	20.1	6.0	20.5	6.2
C6F15V1D1	1.65	10.7	4.2	11.0	5.1
C6F15V1D2	1.59	11.3	4.3	11.8	6.2
C6F15V1D3	1.45	12.1	3.9	12.9	5.8
C6F15V2D1	2.57	9.9	3.9	10.1	4.5
C6F15V2D2	2.46	10.7	4.0	11.0	5.5
C6F15V2D3	2.26	11.7	3.8	12.5	5.4
C6F30V1D1	1.48	9.6	3.4	10.2	5.3
C6F30V1D2	1.56	11.7	2.8	10.8	3.2
C6F30V1D3	1.45	5.8	2.7	6.0	2.8
C6F30V2D1	n/a	n/a	n/a	n/a	n/a
C6F30V2D2	2.25	10.7	2.8	11.2	4.5
C6F30V2D3	2.28	5.7	2.5	5.6	2.4

n/a = not available.

Table 22. 8-ft culvert run data.

Run ID	V_{max} (ft/s)	Fitted RMSE (percent)		Estimated RMSE (percent)	
		Grid	Slice	Grid	Slice
C8F00V1D1	n/a	n/a	n/a	n/a	n/a
C8F00V1D2	n/a	n/a	n/a	n/a	n/a
C8F00V1D3	n/a	n/a	n/a	n/a	n/a
C8F00V2D1	n/a	n/a	n/a	n/a	n/a
C8F00V2D2	n/a	n/a	n/a	n/a	n/a
C8F00V2D3	4.66	18.5	4.4	17.9	4.8
C8F15V1D1	1.58	8.9	7.4	11.4	9.5
C8F15V1D2	n/a	n/a	n/a	n/a	n/a
C8F15V1D3	1.56	9.1	2.2	11.8	8.4
C8F15V2D1	n/a	n/a	n/a	n/a	n/a
C8F15V2D2	n/a	n/a	n/a	n/a	n/a
C8F15V2D3	4.44	8.4	2.8	9.9	5.2
C8F30V1D1	n/a	n/a	n/a	n/a	n/a
C8F30V1D2	n/a	n/a	n/a	n/a	n/a
C8F30V1D3	1.51	6.9	2.0	7.5	4.3
C8F30V2D1	n/a	n/a	n/a	n/a	n/a
C8F30V2D2	n/a	n/a	n/a	n/a	n/a
C8F30V2D3	4.45	4.8	2.7	5.3	2.8

n/a = not available.

APPENDIX B. DATA COLLECTION TECHNIQUES

This appendix provides additional details regarding data collection techniques and instrumentation.

ADV

ADV is an intrusive, remote-sensing technique originally developed for hydrodynamic investigations at U.S. Army Engineer Waterways Experiment Station. The theory is based on the shift in received frequency, that is, the so-called Doppler effect. The device sends out a beam of acoustic waves at a fixed frequency from a transmitter probe. These waves bounce off moving particulate matter in the water, and three acoustic receivers sense the shift in the frequency.

Figure 105 depicts the operation principle of the Doppler measurement technique.[33] The transmit transducer produces periodic short acoustic pulses. As the pulses travel along the beam, microbubbles, suspended sediment, or seeding material scatter a tiny fraction of the acoustic energy. These acoustic echoes are detected by the receive transducers if they originate at the sampling volume defined by the intersection of the transmit and receive beams. The frequency of the echo is Doppler shifted according to the relative motion of the scatters, assumed to be traveling with the velocity of the fluid flow. Orthogonal components can be computed by knowledge of the geometry of the beams. The quality of the measurement is dependent on the presence of scatterers and their behavior within the sampling volume. To ensure that ADV measurements provide an accurate representation of the flow velocity, one should evaluate two additional parameters, the signal to noise ratio and the correlation. Filtering the data using one or both of these parameters can improve the quality of measurement.[34] For more information, the reader is referred to SonTek™ and Precht, et al.[21,35]

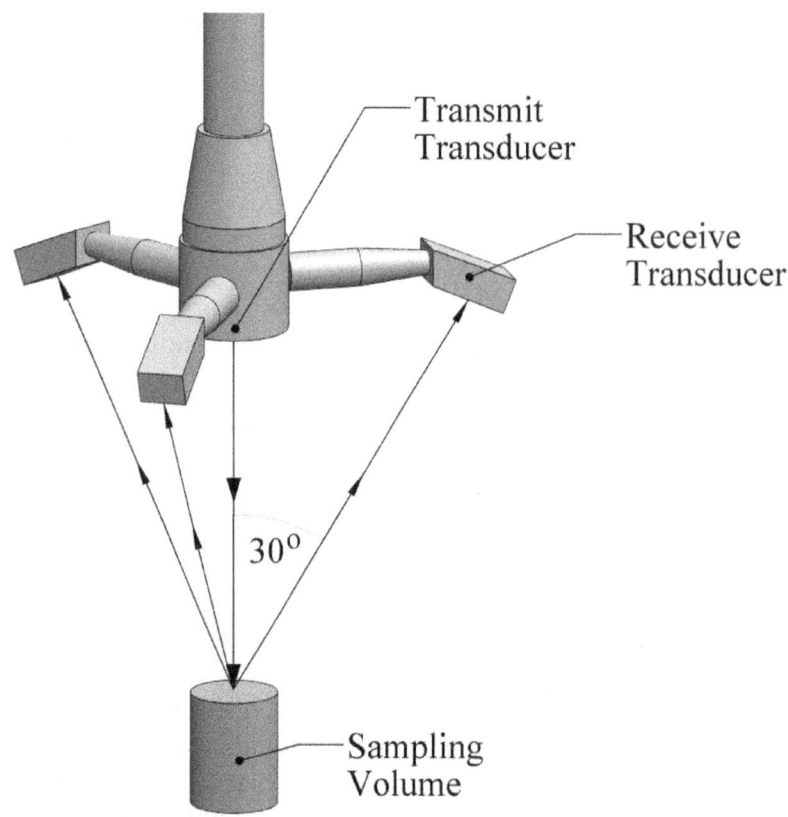

Figure 105. Illustration. Measurement probe.

PIV

PIV is a non-intrusive and whole flow field measuring technique for assessment of the mean and instantaneous velocity vectors within a single plane of interest. PIV in its simplest form consists of a double-pulsed laser with a synchronized camera equipped with a CCD employed to capture particle displacements in successive video frames.[22,36,37] Subsequently, image-processing algorithms are used to arrive at a final velocity distribution with an exceptional spatial resolution. Adoption of a density close to that of a testing medium and an appropriate size of particles are of great importance because they ensure that seeding particles are faithfully following the flow as well as scattering enough incident laser light even with low laser energy.[38]

Sophisticated real-world problems raised the interest among scholars to direct the studies toward 3D PIV with increased temporal and spatial resolution. Reconstructing out-of-plane velocity vectors for highly 3D flows and accounting for traditional PIV perspective errors caused by the imbedded velocity component were major dilemmas that needed to be addressed. Scientists focused on two technical categories: 1) those in which the 3D velocity is calculated from 3D domain (e.g., holographic PIV, tomographic PIV, and scanning PIV), or 2) those in which the 3D velocity is reconstructed from a 2D domain. Because of its long calculation times and costly apparatus, the first category of techniques does not apply to many practical needs. For the latter one, SPIV, dual-plane SPIV, and off-axis SPIV have been successfully developed.[37] In SPIV, two coupled cameras capture the same plane at the same time, but with different off-axis view angles. Both cameras should focus on the same spot in the testing medium and be calibrated properly.[39] Velocity components that are obtained from cross-correlated dewarped images are sufficient to retrieve the third out-of-plane velocity component.[40,41,42]

Human 3D perception of 2D views (binocular vision) is achieved with the coordinated use of both eyes. From a technical point of view, stereovision is the impression of the third spatial dimension (i.e., depth) from two dissimilar views of the same scene. Being inspired by this notion, SPIV was developed using two approaches. In the first approach, translational displacement configuration (figure 106), the disparity is accomplished by having CCD cameras with their optical axis parallel to each other and perpendicular to the object plane. (See references 43 through 46.) In the second approach, angular displacement (figure 107), the camera lens axis subtends an oblique angle to the laser sheet.[41,42]

The translational method offers advantages of convenient mapping, is easy to apply, and produces well-focused images. However, reduced overlap field of view is one of the shortcomings.[46] The angular displacement approach is limited by an upper bound to the off-axis angle subtended by the center of the lens to the center of the region of interest (because of the lens design). Limitations in both approaches motivated the development of an alternative approach.

To address the upper bound off-axis angle restriction in angular displacement systems, the camera's optical axes are aligned neither parallel with each other nor orthogonal to the object plane in an alternative strategy.[41,42] This configuration significantly reduces the out-of-plane velocity component relative error. The angular system similar to that of Willert has been employed in this project.[41]

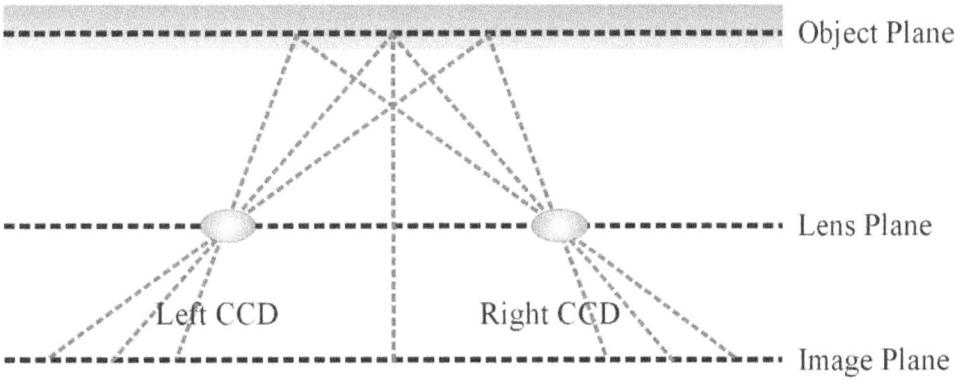

Figure 106. Illustration. Translational configuration.

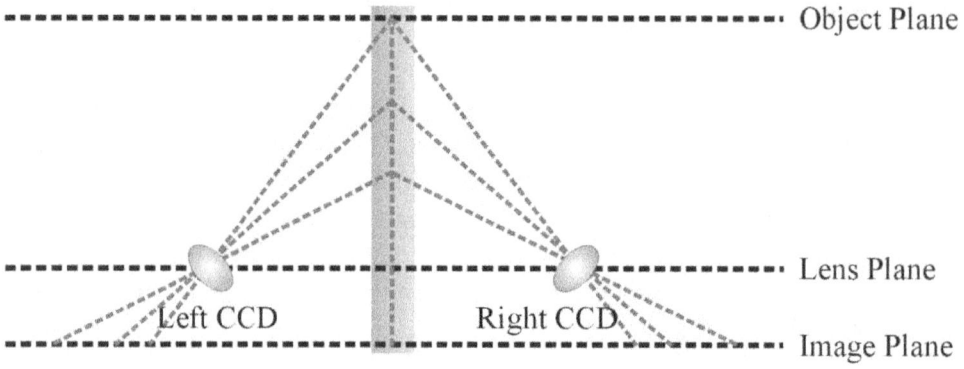

Figure 107. Illustration. Angular configuration.

APPENDIX C. APPLICATION GUIDE

This application guide describes a simplified procedure for assessing fish passage at low flows, recognizing velocity variations in the flow field. The procedure employs existing tools, such as HY-8, for making many of the computations needed for the procedure, including computation of composite roughness, flow depth, and average velocity. FHWA anticipates further modifications to HY-8 that will completely automate the calculations needed for this procedure.

FHWA and others have provided comprehensive guidance for the use of stream simulation to design for fish passage, but as described previously, stream simulation may not be appropriate for all situations. This procedure employs the fish passage path width concept with the simplified design charts developed in this research report.

The first part of this guide describes the basic steps of the process, referring the reader to other standard references for computational details. The second part includes example computations for three embedded culvert design situations. The first two include circular culverts, and the third considers a box culvert.

DESIGN PROCEDURE

The simplified design procedure for assessing fish passage in a culvert is composed of five steps. Most of the steps rely on well-established methodologies described in other references. The design charts provided in this report are used in step 5.

Step 1. Estimate Design Flows

A culvert may be initially sized or assessed using a peak discharge developed using traditional methods. The design flows of interest for fish passage are those that are relevant to the periods when fish require passage through the culvert. As described in detail in HEC 26, this range is bounded by the low passage flow, Q_L, and the high passage flow, Q_H.[1] HEC 26 provides guidance on how those flows are determined, with the first priority being site-specific guidance provided by the appropriate Federal, State, or local regulatory authority.

HEC 26 goes on to explain that in the absence of site-specific guidelines, the Q_H should be defined as either the 10-percent exceedance quantile on the annual flow duration curve or 25 percent of the Q_2, in that order of preference. Similarly, in the absence of site-specific guidance, Q_L should be defined as either the 90-percent exceedance quantile on the annual flow duration curve or the 7-day, 2-year low flow (7Q2), whichever is smaller. Regardless of the source of the estimate, Q_L should not be lower than 1 ft^3/s. More detailed information is provided in HEC 26.

Step 2. Estimate Flow Depth and Composite Roughness

For an embedded culvert, the wetted perimeter of the flow includes two different materials, culvert and stream bed, and therefore two different roughness values. The values for the culvert material can be taken from standard hydraulic references, such as HDS 5, and the values for stream bed materials are available in a variety of references.[18]

HEC 26 provides a summary of several methods appropriate for estimating Manning's n value for bed materials. The method presented in the main report is the Blodgett equation, which performs well for the range of conditions encountered for fish passage design.[1] The Blodgett equation is shown in figure 108.

$$n = \frac{\alpha \, y^{1/6}}{2.25 + 5.23 \, log\left(\dfrac{y}{D_{50}}\right)}$$

Figure 108. Equation. Blodgett's equation for bed roughness.

Where:
n = Manning's roughness, dimensionless.
y = Flow depth, ft.
D_{50} = Median grain size, ft.
α = 0.262 in customary units and 0.319 in metric units.

Once the Manning's roughness values for the culvert wall and bed material are determined, a composite roughness value is needed. The equation recommended in this study is provided in figure 109.

$$n_c = \left[\frac{P_b n_b^{1.5} + P_w n_w^{1.5}}{P_b + P_w}\right]^{2/3}$$

Figure 109. Equation. Composite roughness.

Where:
n_c = Composited n-value for the culvert.
P_b = Wetted perimeter of the bed material in the culvert.
n_b = n-value of the bed material in the culvert.
P_w = Wetted perimeter of the culvert walls above the bed material.
n_w = n-value of the culvert material.

This equation is also used in HDS 5 and recommended in the research work of Tullis.[29] It is also used in HY-8 so that a designer using this tool will only need to enter the individual roughness values, and HY-8 performs the compositing computation.

Step 3. Compute Water Surface Top Width and Average Velocity

The flow profile in the culvert is needed to determine the average velocities throughout the culvert length. The most conservative location to assess from a fish passage perspective is the location in the culvert with the lowest depth because this will be the location with the highest velocity. A computer program such as HY-8 or HEC-RAS may be used for this purpose.

The water surface top width is also required. For a circular conduit, the water surface top width, T, is a function of the culvert diameter, D, the depth of embedment, d_e, and the maximum flow depth, y_{max}, as shown in figure 110. It is calculated using the equation in figure 111.

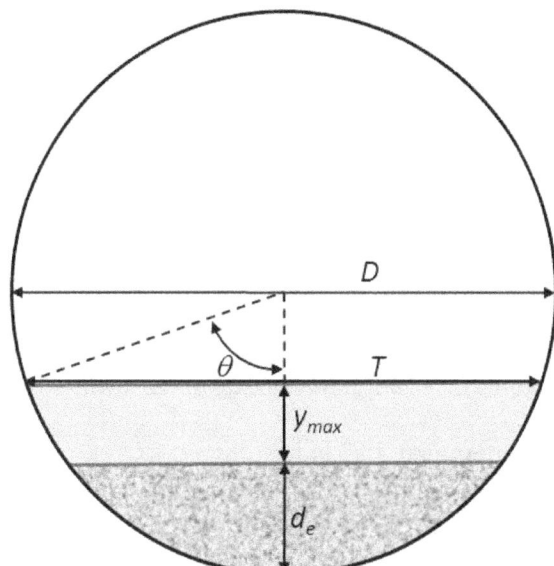

Figure 110. Illustration. Definition sketch for geometric parameters.

$$T = D\sin(\theta)$$

Figure 111. Equation. Water surface top width.

Where:
T = Top width of the water surface, ft.
D = Culvert diameter, ft.
θ = Angle between a line to the water surface at the edge of the pipe from the center of the pipe to the vertical.

$$\theta = a\cos\left(1 - \frac{2(y_{max} + d_e)}{D}\right)$$

Figure 112. Equation. Angle to water surface.

Where:
y_{max} = Maximum flow depth, ft.
d_e = Embedment depth, ft.

Step 4. Find Appropriate Fish Swimming Characteristics

Use of this simplified design procedure requires identification of the required depth and velocity limits for passage for target fish species. The species of interest will likely be specified by the appropriate regulator or by a design team member with expertise in fish biology. In many cases, written guidance may be found in State or local design manuals. HEC 26 provides a national overview of fish biology and swimming capabilities.

The parameters required for design are maximum allowable fish passage velocity, V_f, and minimum depth required for fish passage, y_f, in the culvert for the range of conditions between the low and high passage flows. In some cases, the length of the culvert may also be relevant because fish can only swim against higher velocities for a limited duration before tiring. If the culvert is too long, the fish may not be able to pass the culvert. See HEC 26 for more information on this topic.

Step 5. Determine Fish Passage Path Width

Finally, the available fish passage width, if any, is determined using the method and design charts described in this research report. If the width is insufficient, changing the culvert size, shape, slope, or other options should be evaluated to determine whether a feasible strategy for satisfying the design criteria exists.

Four design curves are available for a range of conditions as follows:

1. Culvert with no embedment and T/D is less than 0.8.

2. Culvert with no embedment and T/D is greater than or equal to 0.8.

3. Culvert with embedment less than 20 percent of the culvert diameter. (Note that HEC 26 recommends a minimum embedment of 30 percent for circular culverts.[1])

4. Culvert with embedment greater than or equal to 20 percent of the culvert diameter.

These design curves are shown graphically in figure 113 and in a tabular format in table 23. They are presented in a dimensionless format where the variables are defined as follows:

- z is the horizontal distance from the culvert centerline (negative sign means moving to the left of the centerline), ft (m).

- B_l is the water surface width from the centerline to the left edge of water (equals $T/2$ in a symmetrical culvert), ft (m).

- V is the vertically averaged velocity at a given location, ft/s (m/s).

- V_a is the average flow velocity in the culvert, ft/s (m/s).

The design curves are applicable to circular, elliptical, pipe arch, and box shapes. The curves were developed for conditions where the flow top width, T, stays the same or increases with increasing water surface elevation. For box culverts, the top width remains the same for all water surface elevations. For circular, elliptical, and pipe arch shapes, the top width increases with increasing water surface elevation until the water surface elevation reaches the spring line of the culvert. Further increases in water surface elevation result in a decrease in flow top width. The design curves may only be used for water surface elevations above the spring line when the water surface top width has decreased 2 percent or less from the maximum top width. Curve 4 is used for all box shapes.

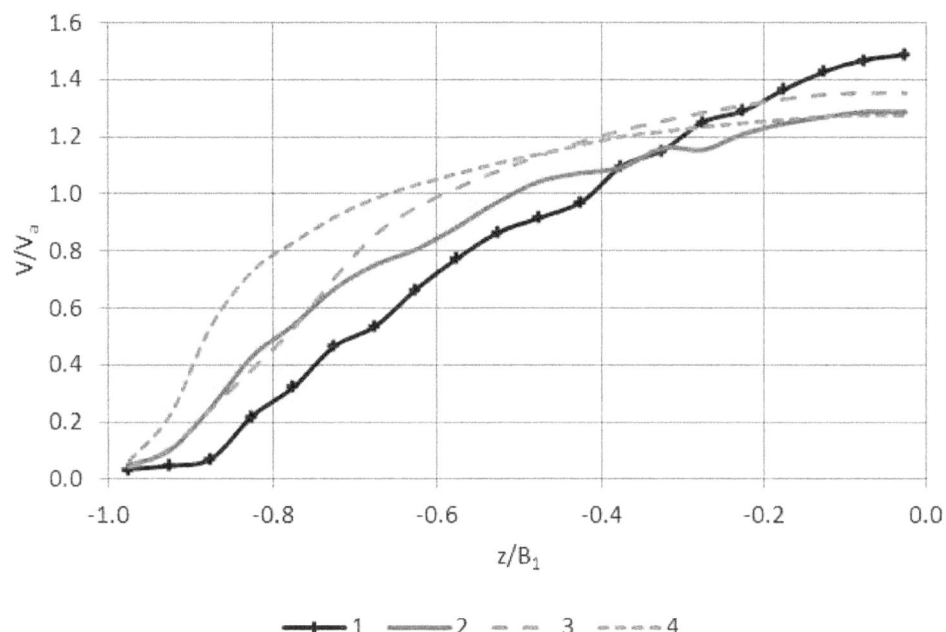

Figure 113. Graph. Vertical slice design curves.

In addition to the limits on water surface elevation for circular, elliptical, and pipe arch culverts, these design curves are limited to the range of hydraulic conditions evaluated in the study. The Froude number, calculated based on the average depth and average velocity in the culvert, should be within the range of 0.1 to 0.5. It is also recommended that these curves only be applied to culverts with spans between 3 and 15 ft, inclusive.

Table 23. Tabular listing of vertical slice design curves.

Slice	z/B_1	V/V_a Curve 1	V/V_a Curve 2	V/V_a Curve 3	V/V_a Curve 4
1	-0.975	0.036	0.048	0.043	0.064
2	-0.925	0.047	0.103	0.108	0.225
3	-0.875	0.072	0.252	0.246	0.537
4	-0.825	0.221	0.430	0.386	0.729
5	-0.775	0.322	0.542	0.529	0.830
6	-0.725	0.465	0.667	0.705	0.917
7	-0.675	0.537	0.751	0.857	0.985
8	-0.625	0.666	0.806	0.951	1.032
9	-0.575	0.774	0.884	1.021	1.071
10	-0.525	0.865	0.974	1.081	1.108
11	-0.475	0.915	1.044	1.134	1.140
12	-0.425	0.972	1.073	1.182	1.172
13	-0.375	1.097	1.092	1.222	1.201
14	-0.325	1.153	1.160	1.256	1.218
15	-0.275	1.253	1.155	1.285	1.235
16	-0.225	1.291	1.211	1.310	1.248
17	-0.175	1.366	1.246	1.332	1.259
18	-0.125	1.428	1.269	1.348	1.270
19	-0.075	1.470	1.287	1.353	1.274
20	-0.025	1.488	1.287	1.353	1.274

EXAMPLE COMPUTATIONS

Example computations for three situations are illustrated. The primary tools used in these examples are the design curves presented in this research report and HY-8 to calculate culvert characteristics.

Example 1. Embedded 6-ft Diameter CMP

For this example, a 6-ft diameter CMP with a length of 53 ft is considered. Using HEC 26 guidance, it is embedded 2 ft, the minimum recommended for a circular conduit. The stream bed material that will be placed in the culvert is a noncohesive material with a D_{50} of 0.944 inches and a D_{84} of 2.6 inches, which is the same material found in the stream. The tailwater channel is approximated by a trapezoidal channel with a bottom width of 5.5 ft and 1:1 side slopes. Both the culvert and the tailwater channel have a longitudinal slope of 0.15 percent.

Step 1. Estimate Design Flows

The high and low passage flows were determined to be 7.8 and 1.0 ft^3/s, respectively, for this hypothetical location. See HEC 26 for more detailed discussion of the passage flows.

Step 2. Estimate Flow Depth and Composite Roughness

Manning's roughness for the culvert is obtained from HDS 5 for a CMP as 0.028. For the bed material, which is the same material in both the stream and within the culvert, Blodgett's equation is used. (See equation in figure 108.) Because the Manning's n for noncohesive bed materials is dependent on depth and depth is dependent on Manning's n, an iterative process is required. Considering the high passage discharge, a trial normal depth of 1 ft is assumed for the tailwater channel. Then, Manning's n is computed as follows (D_{50} is converted to units of feet):

$$n = \frac{0.262 \, y^{\frac{1}{6}}}{2.25 + 5.23 \log\left(\dfrac{y}{D_{50}}\right)} = \frac{0.262 \, (1)^{\frac{1}{6}}}{2.25 + 5.23 \log\left(\dfrac{1}{0.0786}\right)} = \frac{0.262}{8.03} = 0.033$$

This *n* value is entered into HY-8 to represent both the tailwater and bed material Manning's n values. HY-8 is run, and the normal depth is given. This new normal depth is then entered into the Blodgett equation to compute a new Manning's n. This process is repeated until there are no further changes to Manning's n (within two significant figures). In this case, the normal depth was overestimated, but the Manning's n value does not change with the new depth. Therefore, a Manning's roughness value of 0.033 is used for the high passage flow.

The individual values for the culvert and bed material are entered into HY-8 for computation of composite roughness. For this case, the composite roughness is 0.032 for the high passage flow.

For the low passage flow, the same roughness value is used for the CMP, but the bed material value will increase because of the shallower depth. Using the same process, the Manning's roughness for the bed material is 0.041. Using HY-8, the composite value is 0.040.

Step 3. Compute Water Surface Top Width and Average Velocity

Using HY-8 for the high passage flow and the associated roughness value, the flow profile in the culvert is determined to vary from a depth of 0.88 ft at the culvert outlet to 0.89 ft at the culvert inlet. Therefore, the highest velocity in the culvert will occur at the outlet. From HY-8, that velocity is 1.51 ft/s.

The water surface top width is also needed and, for an embedded circular pipe, is computed from the equation in figure 111 with the support of the equation in figure 112. From the latter, the angle between a line to the water surface at the edge of the pipe from the center of the pipe to the vertical is calculated as:

$$\theta = a\cos\left(1 - \frac{2(y_{max} + d_e)}{D}\right) = a\cos\left(1 - \frac{2(0.88 + 2.0)}{6}\right) = 87.7 \ deg\,rees$$

The water surface top width T is calculated from:

$$T = D\sin(\theta) = 6\sin(87.7) = 5.995 \ ft$$

Because the total of the embedment and water surface depth brings the water surface close to the center of the culvert, the water surface top width is close to the culvert span for the high passage flow.

Using HY-8 for the low passage flow and the associated roughness value, the flow profile in the culvert is determined to be close to a constant 0.29 ft at the outlet and inlet. From HY-8, the velocity associated with that depth is 0.59 ft/s.

The water surface top width may be needed and is computed as before:

$$\theta = a\cos\left(1 - \frac{2(y_{max} + d_e)}{D}\right) = a\cos\left(1 - \frac{2(0.29 + 2.0)}{6}\right) = 76.3 \ deg\,rees$$

The water surface top width T is calculated from:

$$T = D\sin(\theta) = 6\sin(76.3) = 5.829 \ ft$$

Step 4. Find Appropriate Fish Swimming Characteristics

For the target fish species identified for this hypothetical case, the maximum velocity is 1.3 ft/s, and the minimum depth is 0.5 ft. At the low passage flow, the velocity meets the requirements for passage, but the depth of 0.29 ft does not. This is a common occurrence for the low passage flow. The recommended design approach to address this situation is to form a v-shaped channel within the culvert that provides the needed depth. See HEC 26 for more information.

Step 5. Determine Fish Passage Path Width

Given the average velocity of 1.51 ft/s under the high passage flow, it appears this culvert is a barrier to passage. The design charts of figure 113 and table 23 are used to determine whether a portion of the culvert includes velocities that are passable.

For the case of a circular culvert with an embedment of 33 percent (2 ft divided by 6 ft), design curve number 4 is appropriate. The ratio of the allowable fish velocity to the average velocity in the culvert is 1.3/1.51 = 0.86. For curve 4 in table 23, this value falls between the values for vertical slice 5 and vertical slice 6. Therefore, the leftmost five slices provide average velocities that are passable by the target fish.

Recall that each slice is one twentieth of one half of the water surface width (not the culvert span). As computed under step 3, the width, T, is 5.995 ft. Therefore, the fish passage width is 5 x (1/20) x (5.995/2) = 0.75 ft. Because this culvert is symmetrical, there is this same passage width on both the left and right edges of the culvert, with velocities sufficiently low that the target fish species can pass at the high passage flow.

The final check is whether there is sufficient depth for the target fish to pass. Inspection of the cross-section graphic in HY-8 shows that the depth is sufficient for passage throughout the estimated fish passage width. Therefore, the total fish passage width in the culvert for the target species is 1.5 ft, even though the average velocity in the culvert exceeds the allowable passage velocity.

Example 2. Embedded 8-ft Diameter CSP

For this example, an 8-ft diameter CSP with a length of 60 ft is considered. Using HEC 26 guidance, it is embedded 2.4 ft. (For circular culverts, 2 ft or 30 percent of the culvert diameter, whichever is larger, is recommended.) The stream bed material that will be placed in the culvert is a noncohesive material with a D_{50} of 2.36 inches and a D_{84} of 6.50 inches, which is the same material found in the stream. The tailwater channel is approximated by a trapezoidal channel with a bottom width of 7.5 ft and 1:1 side slopes. Both the culvert and the tailwater channel have a longitudinal slope of 0.5 percent.

Step 1. Estimate Design Flows

The high and low passage flows were determined to be 16.7 and 1.5 ft³/s, respectively, for this hypothetical location. See HEC 26 for more detailed discussion of the passage flows.

Step 2. Estimate Flow Depth and Composite Roughness

Manning's roughness for the culvert is obtained from HDS 5 for a CSP as 0.035. For the bed material, which is the same material in both the stream and within the culvert, Blodgett's equation is used. (See equation in figure 108.) Because the Manning's n for noncohesive bed materials is dependent on depth and depth is dependent on Manning's n, an iterative process is required. Considering the high passage discharge, a trial normal depth of 1 ft is assumed for the tailwater channel. Then, Manning's n is computed as follows (D_{50} is converted to units of feet):

$$n = \frac{0.262\, y^{1/6}}{2.25 + 5.23\, log\left(\dfrac{y}{D_{50}}\right)} = \frac{0.262\,(1)^{1/6}}{2.25 + 5.23\, log\left(\dfrac{1}{0.197}\right)} = \frac{0.262}{5.94} = 0.044$$

This *n* value is entered into HY-8 to represent both the tailwater and bed material Manning's n values. HY-8 is run, and the normal depth is given. This new normal depth is then entered into the Blodgett equation to compute a new Manning's n. This process is repeated until there are no further changes to Manning's n (within two significant figures). In this case, the normal depth was slightly overestimated, but the Manning's n value does not change with the new depth. Therefore, a Manning's roughness value of 0.044 is used for the high passage flow.

The individual values for the culvert and bed material are entered into HY-8 for computation of composite roughness. For this case, the composite roughness is 0.042 for the high passage flow.

For the low passage flow, the same roughness value is used for the CMP, but the bed material value will increase because of the shallower depth. Using the same process, the Manning's roughness for the bed material is 0.067. Using HY-8, the composite value is 0.064.

Step 3. Compute Water Surface Top Width and Average Velocity

Using HY-8 for the high passage flow and the associated roughness value, the flow profile in the culvert is determined to vary from a depth of 0.96 ft at the culvert outlet to 0.99 ft at the culvert inlet. Therefore, the highest velocity in the culvert will occur at the culvert outlet. From HY-8, that velocity is 2.29 ft/s.

The water surface top width is also needed and, for an embedded circular pipe, is computed from the equation in figure 111 with the support of the equation in figure 112. From the latter, the angle between a line to the water surface at the edge of the pipe from the center of the pipe to the vertical is calculated as:

$$\theta = acos\left(1 - \frac{2(y_{max} + d_e)}{D}\right) = acos\left(1 - \frac{2(0.96 + 2.4)}{8}\right) = 80.8 \; deg\,rees$$

The water surface top width T is calculated from:

$$T = D\,sin(\theta) = 8\,sin(80.8) = 7.90 \; ft$$

Because the total of the embedment and water surface depth brings the water surface close to the center of the culvert, the water surface top width is close to the culvert span for the high passage flow.

Using HY-8 for the low passage flow and the associated roughness value, the flow profile in the culvert is determined to be close to a constant 0.29 ft at the outlet and inlet. From HY-8, the velocity associated with that depth is 0.71 ft/s.

The water surface top width may be needed and is computed as before:

$$\theta = acos\left(1 - \frac{2(y_{max} + d_e)}{D}\right) = acos\left(1 - \frac{2(0.29 + 2.4)}{8}\right) = 70.9 \; deg\,rees$$

The water surface top width T is calculated from:

$$T = D\,sin(\theta) = 8\,sin(70.9) = 7.56 \; ft$$

Step 4. Find Appropriate Fish Swimming Characteristics

For the target fish species identified for this hypothetical case, the maximum velocity is 1.3 ft/s, and the minimum depth is 0.5 ft. At the low passage flow, the velocity meets the requirements for passage, but the depth of 0.29 ft does not. This is a common occurrence for the low passage flow. The recommended design approach to address this situation is to form a v-shaped channel within the culvert that provides the needed depth. See HEC 26 for more information.

114

Step 5. Determine Fish Passage Path Width

Given the average velocity of 2.29 ft/s under the high passage flow, it appears this culvert is a barrier to passage. The design charts of figure 113 and table 23 are used to determine whether a portion of the culvert includes velocities that are passable.

For the case of a circular culvert with an embedment of 30 percent (2.4 ft divided by 8 ft) design curve number 4 is appropriate. The ratio of the allowable fish velocity to the average velocity in the culvert is 1.3/2.29 = 0.57. For curve 4 in table 23, this value falls between the values for vertical slice 3 and vertical slice 4. Therefore, the leftmost three slices provide average velocities that are passable by the target fish.

Recall that each slice is one twentieth of one half of the water surface width (not the culvert span). As computed under step 3, the width, T, is 7.90 ft. Therefore, the fish passage width is 3 x (1/20) x (7.90/2) = 0.59 ft. Because this culvert is symmetrical, there is this same passage width on both the left and right edges of the culvert, with velocities sufficiently low that the target fish species can pass at the high passage flow.

The final check is whether there is sufficient depth for the target fish to pass. Inspection of the cross-section graphic in HY-8 shows that the depth is sufficient for passage throughout the estimated fish passage width. Therefore, the total fish passage width in the culvert for the target species is 1.18 ft, even though the average velocity in the culvert exceeds the allowable passage velocity.

Example 3. Embedded 8 ft x 8 ft Concrete Box Culvert

For this example, an 8- x 8-ft concrete box culvert (CBC) with a length of 60 ft is considered. Using HEC 26 guidance, it is embedded 2 ft, the minimum recommended for a box culvert. The stream bed material that will be placed in the culvert is a noncohesive material with a D_{50} of 2.36 inches and a D_{84} of 6.50 inches, which is the same material found in the stream. The tailwater channel is approximated by a trapezoidal channel with a bottom width of 7.5 ft and 1:1 side slopes. Both the culvert and the tailwater channel have a longitudinal slope of 0.5 percent.

Step 1. Estimate Design Flows

The high and low passage flows were determined to be 16.7 and 1.5 ft³/s, respectively, for this hypothetical location. See HEC 26 for more detailed discussion of the passage flows.

Step 2. Estimate Flow Depth and Composite Roughness

Manning's roughness for the culvert is obtained from HDS 5 for a CBC as 0.014. For the bed material, which is the same material in both the stream and within the culvert, Blodgett's equation is used. (See equation in figure 108.) Because the Manning's n for noncohesive bed materials is dependent on depth and depth is dependent on Manning's n, an iterative process is required. Considering the high passage discharge, a trial normal depth of 1 ft is assumed for the tailwater channel. Then Manning's n is computed as follows (D_{50} is converted to units of feet):

$$n = \frac{0.262 \, y^{1/6}}{2.25 + 5.23 \log\left(\dfrac{y}{D_{50}}\right)} = \frac{0.262 \, (1)^{1/6}}{2.25 + 5.23 \log\left(\dfrac{1}{0.197}\right)} = \frac{0.262}{5.94} = 0.044$$

This n value is entered into HY-8 to represent both the tailwater and bed material Manning's n values. HY-8 is run, and the normal depth is given. This new normal depth is then entered into the Blodgett equation to compute a new Manning's n. This process is repeated until there are no further changes to Manning's n (within two significant figures). In this case, the normal depth was overestimated, but the Manning's n value does not change with the new depth. Therefore, a Manning's roughness value of 0.044 is used for the high passage flow.

The individual values for the culvert and bed material are entered into HY-8 for computation of composite roughness. For this case, the composite roughness is 0.039 for the high passage flow.

For the low passage flow, the same roughness value is used for the CMP, but the bed material value will increase because of the shallower depth. Using the same process, the Manning's roughness for the bed material is 0.067. Using HY-8, the composite value is 0.064.

Step 3. Compute Water Surface Top Width and Average Velocity

Using HY-8 for the high passage flow and the associated roughness value, the flow profile in the culvert is determined to vary from a depth of 0.96 ft at the outlet to 0.94 ft at the inlet. Therefore, the highest velocity in the culvert will occur at the inlet. From HY-8, that velocity is 2.21 ft/s.

The water surface top width for a box culvert is always the span of the culvert, which in this case is 8 ft.

Using HY-8 for the low passage flow and the associated roughness value, the flow profile in the culvert is determined to be 0.29 ft at the outlet and 0.28 ft at the inlet. From HY-8, the velocity associated with that depth is 0.67 ft/s.

The water surface top width may be needed and is 8 ft, as it was for the high passage flow.

Step 4. Find Appropriate Fish Swimming Characteristics

For the target fish species identified for this hypothetical case, the maximum velocity is 1.3 ft/s, and the minimum depth is 0.5 ft. At the low passage flow, the velocity meets the requirements for passage, but the depth of 0.28 ft does not. This is a common occurrence for the low passage flow. The recommended design approach to address this situation is to form a v-shaped channel within the culvert that provides the needed depth. See HEC 26 for more information.

Step 5. Determine Fish Passage Path Width

Given the average velocity of 2.21 ft/s under the high passage flow, it appears this culvert is a barrier to passage. For all box culverts, design curve 4 from figure 113 and table 23 is used to determine whether a portion of the culvert includes velocities that are passable.

The ratio of the allowable fish velocity to the average velocity in the culvert is 1.3/2.21 = 0.59. For curve 4 in table 23, this value falls between the values for vertical slice 3 and vertical slice 4. Therefore, the leftmost three slices provide average velocities that are passable by the target fish.

Recall that each slice is one twentieth of one half of the water surface width. As computed under step 3, the width, T, is 8.0 ft. Therefore, the fish passage width is 3 x (1/20) x (8/2) = 0.60 ft. Because this culvert is symmetrical, there is this same passage width on both the left and right edges of the culvert, with velocities sufficiently low that the target fish species can pass at the high passage flow.

The final check is whether there is sufficient depth for the target fish to pass. Because the depth is constant across the entire span, the depth is sufficient for passage throughout the estimated fish passage width. Therefore, the total fish passage width in the culvert for the target species is 1.2 ft, even though the average velocity in the culvert exceeds the allowable passage velocity.

ACKNOWLEDGMENTS

This research was conceived and developed to meet the needs of the contributors to a pooled-fund effort (TPF-5(164)). The States that contributed funding were Alaska, Georgia, Maryland, Michigan, Minnesota, Vermont, and Wisconsin. It was supported by the FHWA Hydraulics Research and Development Program with Contract No. DTFH61-11-D-00010 by Genex Systems and the Argonne National Laboratory.

The Maryland State Highway Administration was the lead State working with the FHWA to develop the concepts for the study. Dr. Fred Chang was involved in the preliminary conceptual discussions that established the research priorities and analytical frameworks as noted in the text. Corrugated pipe used in the study was donated by Contech® at the request of the Maryland State Highway Administration.

Roger Kilgore developed the proposed design equations and provided data analyses and technical editing services.

REFERENCES

1. Kilgore, R.T., B.S. Bergendahl, and R.H. Hotchkiss (2010). *Aquatic Organism Passage Design Guidelines for Culverts*, Hydraulic Engineering Circular, Number 26 (HEC 26), Federal Highway Administration, FHWA-HIF-11-0008.

2. Forest Service Stream Simulation Working Group (2008). *Stream Simulation: An Ecological Approach to Providing Aquatic Organism Passage at Road-Stream Crossings*, National Technology and Development Program, San Dimas, CA, United States Forest Service, May.

3. Hotchkiss, R.H. and C.M. Frei (2007). *Design for Fish Passage and Road-Stream Crossings: Synthesis Report*, Federal Highway Administration.

4. Puertas, J., L. Pena, and T. Teijeiro (2004). "An Experimental Approach to the Hydraulics of Vertical Slot Fishways, *Journal of Hydraulic Engineering, 130*(1).

5. Magura, C.R. (2007). *Hydraulic Characteristics of Embedded Circular Culverts*, Master's Thesis, University of Manitoba, September.

6. Barber, M.E. and R.C. Downs (1996). *Investigation of Culvert Hydraulics Related to Juvenile Fish Passage*, Rep. No. WA-RD 388.2, Washington State Transportation Center (TRAC).

7. Roberson, J.A. and C.T. Crowe (1990). *Engineering Fluid Mechanics*, Houghton Mifflin Company, Boston, MA.

8. Goring, D.G. (1997). "Modeling the Distribution of Velocity in a River Cross Section," *New Zealand Journal of Marine and Freshwater Research, 31.*

9. House, M.R., M.R. Pyles, and D. White (2005). "Velocity Distribution in Streambed Simulation Culverts Used for Fish Passage," *Journal of the American Water Resources Association*, February.

10. Blank, M.D., J.E. Cahoon, and T. McMahon (2008). *Advanced Studies of Fish Passage Through Culverts: 1-D and 3-D Hydraulic Modeling of Velocity, Fish Expenditure and a New Barrier Assessment Method*, Department of Civil Engineering and Ecology, Montana State University.

11. Pati, V.V.R. (2010). *CFD Modeling and Analysis of Flow Through Culverts*, M.S. Thesis, Northern Illinois University.

12. Haque, M.M., G. Constantinescu, and L. Weber (2007). "Validation of a 3D RANS Model to Predict Flow and Stratification Effects Related to Fish Passage at Hydropower Dams," *Journal of Hydraulic Research, 45*(6).

13. Olsen, N.R.B. and S. Stokseth (1995). "Three-dimensional Numerical Modeling of Water Flow in a River with Large Bed Roughness," *Journal of Hydraulic Research, 33.*

14. Hodskinson, A. and R.I. Ferguson (1998). "Numerical Modeling of Separated Flow in River Bends: Model Testing and Experimental Investigation of Geometric Controls on the Extent of Flow Separation at the Concave Bank," *Hydrological Processes, 12.*

15. Lane, S.N. and K.S. Richards (1998). "High Resolution, Two-dimensional Spatial Modeling of Flow Processes in a Multi-thread Channel," *Hydrological Processes, 12.*

16. Nicholas, A.P. and G.H. Sambrooke-Smith (1999). "Numerical Simulation of Three-dimensional Flow Hydraulics in a Braided River," *Hydrological Processes, 13.*

17. Khan, L.A., E.W. Roy, and M. Rashid (2008). "Computational Fluid Dynamics Modeling of Forebay Hydrodynamics Created by a Floating Juvenile Fish Collection Facility at the Upper Baker River Dam, Washington," *River Research and Application, 24.*

18. Schall, J.D., P.L. Thompson, S.M. Zerges, R.T. Kilgore, and J.L. Morris (2012). *Hydraulic Design of Highway Culverts*, Hydraulic Design Series Number 5 (HDS 5), Federal Highway Administration, FHWA-HIF-12-026.

19. Powers, P.D., K. Bates, T. Burns, B. Gowen, and R. Whitney (1997). *Culvert Hydraulics Related to Upstream Juvenile Salmon Passage*, Lands and Restoration Services Program, Washington Department of Fish and Wildlife.

20. Gende, S.M., R.T. Edwards, M.F. Willson, and M.S. Wipfli (2002). "Pacific Salmon in Aquatic and Terrestrial Ecosystems," *Bioscience, 52.*

21. SonTek, (2001). *Sontek/YSI ADV Field/Hydra Acoustic Doppler Velocimeter (Field) Technical Documentation*, Sontek/YSI, San Diego, USA, Cambridge University Press, New York.

22. Adrian, R.J. (1991). "Particle-imaging Techniques for Experimental Fluid Mechanics." *Annual Review of Fluid Mechanics, 23.*

23. Hirt, C.W. and B.D. Nichols (1981). "Volume of Fluid (VOF) Method for the Dynamics of Free Boundaries," *Journal of Computational Physics, 39.*

24. Venkata, S.L. (2011). *Computational Fluid Dynamics Modeling of Flow Through Culverts*, Master's Thesis, Northern Illinois University.

25. Nicholas, A.P. (2001). "Computational Fluid Dynamics Modeling of Boundary Roughness in Gravel-bed Rivers: An Investigation of the Effects of Random Variability in Bed Elevation," *Earth Surface Processes and Landforms, 26.*

26. Lottes, S.A. (2011). *Computational Fluid Dynamics Modeling of Flow Through Culverts*, 2011 Quarter 4 Progress Report, Transportation Research and Analysis Computing Center, Argonne National Laboratory.

27. CD-adapco (2011). User Guide STAR-CCM+ Version 6.04.014.

28. Brunner, G.W. (2010). *HEC-RAS, River Analysis System Hydraulic Reference Manual,* USACE, CPD-69.

29. Tullis, B. (2012). *Hydraulic Loss Coefficients for Culverts,* NCHRP Report 734.

30. Chiu, C. and J. Chiou (1986). "Structure of 3-D Flow in Rectangular Open Channels," *ASCE Journal of Hydraulic Engineering, 112*(11).

31. Chiu, C. (1989). "Velocity Distribution in Open Channel Flow," *ASCE Journal of Hydraulic Engineering, 115*(5).

32. Zhai, Y. (2012). *CFD Modeling of Fish Passage in Large Culverts and Assistance for Culvert Design with Fish Passage*, Ph.D. Dissertation, University of Nebraska, Lincoln, July.

33. Lohrmann, A. and R. Cabera (1994). "New Acoustic Meter for Measuring 3D Laboratory Flows," *Journal of Hydraulic Engineering, 120*(3).

34. Wahl, T.L. (2000). "Analyzing ADV Data Using WinADV," 2000 Joint Conference on Water Resources Engineering and Water Resources Planning and Management, July 30–August 2, 2000, Minneapolis, Minnesota.

35. Precht, E., F. Janssen, and M. Huettel (2006). "Nearbottom Performance of the Acoustic Doppler Velocimeter (ADV)—a Comparative Study," *Aquatic Ecology, 40*(33).

36. Raffel, M., C. Willert, S. Werely, and J. Kompenhans (2007). "*Particle Image Velocimetry: A Practical Guide*," Springer-Verlag, New York.

37. Adrian, R.J. and J. Westerweel (2011). "*Particle Image Velocimetry: A Practical Guide*," Cambridge University Press, New York.

38. Melling, A. (1997). "Tracer Particles and Seeding for Particle Image Velocimetry," *Measurement Science and Technology, 8*(12).

39. Bjorkquist, D.C. (2002). "Stereoscopic PIV Calibration Verification," Proceedings of the 11th International Symposium on Application of Laser Techniques to Fluid Mechanics, Lisbon, Portugal.

40. Keane, R.D. and R.J. Adrian (1992). "Theory of Cross-correlation Analysis of PIV Images," *Applied Science Research, 49*(3).

41. Willert, C. (1997). "Stereoscopic Digital Particle Image Velocimetry for Applications in Wind Tunnel Flows," *Measurement Science and Technology, 8*(12).

42. Prasad, A.K. (2000). "Stereoscopic Particle Image Velocimetry," *Experiments in Fluids, 29*(2).

43. Jacquot, P. and P.K. Rastogi (1981). "Influence of Out-of-plane Deformation and its Elimination in White-light Speckle Photography," *Optics and Lasers in Engineering*, *2*(1).

44. Arroyo, M.P. and C.A. Greated (1991). "Stereoscopic Particle Image Velocimetry," *Measurement Science and Technology*, *2*(12).

45. Prasad, A.K. and R.J. Adrian (1993). "Stereoscopic Particle Image Velocimetry Applied to the Liquid Flows," *Experiments in Fluids*, *15*(1).

46. Gauthier, V. and M.L. Riethmuller (1988). "Application of PIDV to Complex Flows: Measurement of the Third component," VKI Lectures Series on Particle Image Displacement Velocimetry, Brussels.

www.ingramcontent.com/pod-product-compliance
Lightning Source LLC
Chambersburg PA
CBHW080641180526
45168CB00008B/3251